*Evolution and the Theory of Games*

# *Evolution and the Theory of Games*

JOHN MAYNARD SMITH

***Professor of Biology, University of Sussex***

CAMBRIDGE UNIVERSITY PRESS
Cambridge, New York, Melbourne, Madrid, Cape Town,
Singapore, São Paulo, Delhi, Mexico City

Cambridge University Press
The Edinburgh Building, Cambridge CB2 8RU, UK

Published in the United States of America by Cambridge University Press, New York

www.cambridge.org
Information on this title: www.cambridge.org/9780521288842



First published 1982
18th printing 2012

*A catalogue record for this publication is available from the British Library*

*Library of Congress Cataloguing in Publication Data*
Smith, John Maynard
Evolution and the theory of games.
1. Evolution - Mathematical models. 2. Game theory. I. Title
575′.01′5193 QH366.2

ISBN 978-0-521-24673-6 Hardback
ISBN 978-0-521-28884-2 Paperback

# *Contents*

# *Preface*

The last decade has seen a steady increase in the application of concepts from the theory of games to the study of evolution. Fields as diverse as sex ratio theory, animal distribution, contest behaviour and reciprocal altruism have contributed to what is now emerging as a universal way of thinking about phenotypic evolution. This book attempts to present these ideas in a coherent form. It is addressed primarily to biologists. I have therefore been more concerned to explain and to illustrate how the theory can be applied to biological problems than to present formal mathematical proofs – a task for which I am, in any case, ill equipped. Some idea of how the mathematical side of the subject has developed is given in the appendixes.

I hope the book will also be of some interest to game theorists. Paradoxically, it has turned out that game theory is more readily applied to biology than to the field of economic behaviour for which it was originally designed. There are two reasons for this. First, the theory requires that the values of different outcomes (for example, financial rewards, the risks of death and the pleasures of a clear conscience) be measured on a single scale. In human applications, this measure is provided by 'utility' – a somewhat artificial and uncomfortable concept: in biology, Darwinian fitness provides a natural and genuinely one-dimensional scale. Secondly, and more importantly, in seeking the solution of a game, the concept of human rationality is replaced by that of evolutionary stability. The advantage here is that there are good theoretical reasons to expect populations to evolve to stable states, whereas there are grounds for doubting whether human beings always behave rationally.

I have been greatly helped in thinking about evolutionary game theory by my colleagues at the University of Sussex, particularly Brian and Deborah Charlesworth and Paul Harvey. I owe a special debt to Peter Hammerstein, who has helped me to understand some theoretical questions more clearly. The manuscript has been read, in whole or in part, by Jim Bull, Eric Charnov, John Haigh, Peter Hammerstein, Susan Riechert and Siewert Rohwer, all of whom have helped to eliminate errors and ambiguities. Finally it is a pleasure to acknowledge the help of Sheila Laurence, in typing the manuscript and in many other ways.

*November 1981* J. M. S

# 1 *Introduction*

This book is about a method of modelling evolution, rather than about any specific problem to which the method can be applied. In this chapter, I discuss the range of application of the method and some of the limitations, and, more generally, the role of models in science.

Evolutionary game theory is a way of thinking about evolution at the phenotypic level when the fitnesses of particular phenotypes depend on their frequencies in the population. Compare, for example, the evolution of wing form in soaring birds and of dispersal behaviour in the same birds. To understand wing form it would be necessary to know about the atmospheric conditions in which the birds live and about the way in which lift and drag forces vary with wing shape. One would also have to take into account the constraints imposed by the fact that birds' wings are made of feathers – the constraints would be different for a bat or a pterosaur. It would not be necessary, however, to allow for the behaviour of other members of the population. In contrast, the evolution of dispersal depends critically on how other conspecifics are behaving, because dispersal is concerned with finding suitable mates, avoiding competition for resources, joint protection against predators, and so on.

In the case of wing form, then, we want to understand why selection has favoured particular phenotypes. The appropriate mathematical tool is optimisation theory. We are faced with the problem of deciding what particular features (e.g. a high lift:drag ratio, a small turning circle) contribute to fitness, but not with the special difficulties which arise when success depends on what others are doing. It is in the latter context that game theory becomes relevant.

The theory of games was first formalised by Von Neumann & Morgenstern (1953) in reference to human economic behaviour. Since that time, the theory has undergone extensive development;

Luce & Raiffa (1957) give an excellent introduction. Sensibly enough, a central assumption of classical game theory is that the players will behave rationally, and according to some criterion of self-interest. Such an assumption would clearly be out of place in an evolutionary context. Instead, the criterion of rationality is replaced by that of population dynamics and stability, and the criterion of self-interest by Darwinian fitness. The central assumptions of evolutionary game theory are set out in Chapter 2. They lead to a new type of 'solution' to a game, the 'evolutionarily stable strategy' or ESS.

Game theory concepts were first explicitly applied in evolutionary biology by Lewontin (1961). His approach, however, was to picture a species as playing a game against nature, and to seek strategies which minimised the probability of extinction. A similar line has been taken by Slobodkin & Rapoport (1974). In contrast, here we picture members of a population as playing games against each other, and consider the population dynamics and equilibria which can arise. This method of thinking was foreshadowed by Fisher (1930), before the birth of game theory, in his ideas about the evolution of the sex ratio and about sexual selection (see p. 43). The first explicit use of game theory terminology in this context was by Hamilton (1967), who sought for an 'unbeatable strategy' for the sex ratio when there is local competition for mates. Hamilton's 'unbeatable strategy' is essentially the same as an ESS as defined by Maynard Smith & Price (1973).

Most of the applications of evolutionary game theory in this book are directed towards animal contests. The other main area so far has been the problem of sexual allocation – i.e. the sex ratio, parental investment, resource allocation in hermaphrodites etc. I have said rather little on those topics because they are treated at length in a book in preparation by Dr Eric Charnov (1982). Other applications include interspecific competition for resources (Lawlor & Maynard Smith, 1976), animal dispersal (Hamilton & May, 1977), and plant growth and reproduction (Mirmirani & Oster, 1978).

The plan of the book is as follows. In Chapter 2 I describe the basic method whereby game theory can be applied to evolutionary problems. In fact, two different models are considered. The first is that originally proposed by Maynard Smith & Price (1973) to analyse pairwise contests between animals. Although often appropriate for the analysis of fighting behaviour, this model is too restrictive to apply to all cases where fitnesses are frequency-dependent. A second, extended, model is therefore described which can be used when an individual interacts, not with a single opponent at a time, but with some group of other individuals, or with some average property of the population as a whole.

Chapters 3 to 5 deal with other mainly theoretical issues. Chapter 3 analyses the 'war of attrition', whose characteristic feature is that animals can choose from a continuously distributed set of strategies, rather than from a set of discrete alternatives. In Chapter 4, I consider the relationship between game theory models and those of population genetics, and in Chapter 5, the relation between evolution and learning.

Chapters 6 to 10 are concerned with applying the theoretical ideas to field data. My aim here has been to indicate as clearly as possible the different kinds of selective explanation of behaviour that are possible and the kinds of information which are needed if we are to distinguish between them. For most of the examples discussed there are important questions still to be answered. The game-theoretic approach, however, does provide a framework within which a wide range of phenomena, from egg-trading to anisogamy, can be discussed. Perhaps more important, it draws attention to the need for particular kinds of data. In return, the field data raise theoretical problems which have yet to be solved.

The last three chapters are more speculative. Chapter 11 is concerned with how game theory might be applied to the evolution of life history strategies. The particular model put forward, suggested by the evolution of polygynous mammals, is of a rather special and limited kind, but may encourage others to attempt a more general treatment. In Chapter 12, I discuss what may be the most difficult theoretical issue in evolutionary game theory – the transfer of information during contests. Territorial behaviour is discussed in this chapter, because of the theoretical possibility that information transfer will be favoured by selection when the resource being contested is divisible. Finally, Chapter 13 discusses the evolution of cooperation in a game-theoretic context.

The rest of this introductory chapter discusses some more general issues concerned with the application of game theory to evolution.

Those with no taste for philosophical arguments are advised to skip this section, or to treat it as a postscript rather than an introduction; the rest of the book should make sense without it.

First, a word has to be said as to why one should use a game theory model when a classical population genetics model more precisely represents biological reality. The answer is that the two types of model are useful in different circumstances. When thinking about the evolution either of wing shape or of dispersal behaviour it is most unlikely that one would have any detailed knowledge of the genetic basis of variation in the trait. It would, however, be reasonable to suppose that there is some additive genetic variance, because artificial selection experiments have almost always revealed such variance in outbred sexual populations for any selected trait (for a rare exception, see Maynard Smith & Sondhi, 1960). The basic assumption of evolutionary game theory – that like begets like – corresponds to what we actually know about heredity in most cases. To analyse a more detailed genetic model would be out of place. For example, it is relevant to the evolution of wing form that the shape which generates a given lift for the minimum induced drag is an elliptical one. If someone were to say 'Maybe, but how do you know that a bird with an elliptical wing is not a genetic heterozygote which cannot breed true?', he would rightly be regarded as unreasonable.

There are, of course, contests in which population genetic models become necessary. These are discussed in more detail in Chapter 4. Essentially, they are cases in which the centre of interest concerns the genetic variability of the population. Although game theory can sometimes point to situations in which genetic polymorphism can be maintained by frequency-dependent selection, such cases call for proper genetic analysis. Essentially, game theory models are appropriate when one wants to know what phenotypes will evolve, and when it is reasonable to assume the presence of additive genetic variance. Rather surprisingly, game theory methods have proved to be particularly effective in analysing phenotypes (e.g. sex ratio, resource allocation to male and female functions in hermaphrodites) which are themselves relevant to sexual reproduction; all that is required is that the phenotype itself be heritable. The point is also discussed in Chapter 4.

Two further criticisms which can be made of optimisation and game theory models are, first, that it is misleading to think of animals optimising and, secondly, that in any case animals are constrained developmentally and hence unable to reach an optimum. On the first point, optimisation models are certainly misleading if they lead people to think that animals consciously optimise their fitness; they are still more misleading if they lead people to suppose that natural selection leads to the evolution of characteristics which are optimal for the survival of the species. But there is no reason why the models should be so interpreted. An analogy with physical theory should make this point clear. When calculating the path of a ray of light between two points, $A$ and $B$, after reflection or refraction, it is sometimes convenient to make use of the fact that the light follows that path which minimises the time taken to reach $B$. It is a simple consequence of the laws of physics that this should be so; no-one supposes that the ray of light setting out from $A$ calculates the quickest route to $B$. Similarly, it can be a simple consequence of the laws of population genetics that, at equilibrium, certain quantities are maximised. If so, it is simplest to find the equilibrium state by performing the maximisation. Nothing is implied about intention, and nothing is asserted about whether or not the equilibrium state will favour species survival.

On the subject of developmental constraints (see, for example, Gould & Lewontin, 1979), I think there is some misunderstanding. Whenever an optimisation or game-theoretic analysis is performed, an essential feature of the analysis is a specification of the set of possible phenotypes from among which the optimum is to be found. This specification is identical to a description of developmental constraints. I could reasonably claim that by introducing game theory methods I have drawn attention to developmental constraints by insisting that they be specified. Rather than make this claim, I will instead admit that although I can see no theoretical justification for Gould & Lewontin's criticism of the 'adaptationist programme', I can see some practical force to it. This is that, in practice, too much effort is put into seeking an optimum and not enough into defining the phenotype set. In the Hawk–Dove game (p. 11), for example, considerable sophistication has been devoted to analysing the game, but the strategy set is ridiculously naïve.

My reply to this complaint would be that it wrongly identifies the purpose of the Hawk–Dove game, which is not to represent any specific animal example, but to reveal the logical possibilities (for example, the likelihood of mixed strategies) inherent in all contest situations. When confronted with specific cases, much more care must be taken in establishing the strategy set. It is interesting, as an example, that in analysing competition between female digger wasps (p. 74), Brockmann, Grafen & Dawkins (1979) were at first unsuccessful because they wrongly determined the alternative strategies available to the wasps.

There is, however, a wider conflict between the developmental and the evolutionary points of view. After the publication of Darwin's *Origin of Species*, but before the general acceptance of Weismann's views, problems of evolution and development were inextricably bound up with one another. One consequence of Weismann's concept of the separation of germ line and soma was to make it possible to understand genetics, and hence evolution, without understanding development. In the short run this was an immensely valuable contribution, because the problems of heredity proved to be soluble, whereas those of development apparently were not. The long-term consequences have been less happy, because most biologists have been led to suppose either that the problems of development are not worth bothering with, or that they can be solved by a simple extension of the molecular biology approach which is being so triumphant in genetics.

My own view is that development remains one of the most important problems of biology, and that we shall need new concepts before we can understand it. It is comforting, meanwhile, that Weismann was right. We can progress towards understanding the evolution of adaptations without understanding how the relevant structures develop. Hence, if the complaint against the 'adaptationist programme' is that it distracts attention from developmental biology, I have some sympathy. Development is important and little understood, and ought to be studied. If, however, the complaint is that adaptation *cannot* (rather than ought not to) be studied without an understanding of developmental constraints, I am much less ready to agree.

The disagreement, if there is one, is empirical rather than theoretical – it is a disagreement about what the world is like. Thus, I am sure, Gould and Lewontin would agree with me that natural selection does bring about some degree of adaptive fit between organisms and their environments, and I would agree with them that there are limits to the kinds of organisms which can develop. We may disagree, though, about the relative importance of these two statements. Suppose, for example, that only two kinds of wings could ever develop – rectangular and triangular. Natural selection would probably favour the former in vultures and the latter in falcons. But if one asked 'Why are birds' wings the shapes they are?', the answer would have to be couched primarily in terms of developmental constraints. If, on the other hand, almost any shape of wing can develop, then the actual shape, down to its finest details, may be explicable in selective terms.

Biologists differ about which of these pictures is nearer the truth. My own position is intermediate. Clearly, not all variations are equally likely for a given species. This fact was well understood by Darwin, and was familiar to me when I was an undergraduate under the term 'Vavilov's law of homologous variation' (Spurway, 1949; Maynard Smith, 1958). In some cases, the possible range of phenotypic variation may be quite sharply circumscribed; for example, Raup (1966) has shown that the shapes of gastropod shells can be described by a single mathematical expression, with only three parameters free to vary. Further, the processes of development seem to be remarkably conservative in evolution, so that the evolution of legs, wings and flippers among the mammals has been achieved by varying the relative sizes and, to some extent, numbers of parts rather than by altering the basic pattern, or bauplan.

It follows from this that, when thinking about the evolution of any particular group, we must keep in mind the constraints which development is likely to place on variation. Looking at existing mammals, however, makes it clear that the constraint of maintaining a particular basic structure does not prevent the evolution of an extraordinary range of functional adaptations. It would be a mistake to take a religious attitude towards bauplans, or to regard them as revealing some universal laws of form. Our ancestors first evolved a notochord, segmented muscles and two pairs of fins as adaptations for swimming, and not because they were conforming to a law of form. As Darwin remarked in the *Origin*, the 'Unity of Type' is important, but it is subordinate to the 'conditions of existence', because the 'Type' was once an organism which evolved to meet particular conditions of existence.

An obvious weakness of the game-theoretic approach to evolution is that it places great emphasis on equilibrium states, whereas evolution is a process of continuous, or at least periodic, change. The same criticism can be levelled at the emphasis on equilibria in population genetics. It is, of course, mathematically easier to analyse equilibria than trajectories of change. There are, however, two situations in which game theory models force us to think about change as well as constancy. The first is that a game may not have an ESS, and hence the population cycles indefinitely. On the whole, symmetrical games with no ESS seem biologically rather implausible. They necessarily imply more than two pure strategies (see Appendix D), and usually have the property that $A$ beats $B$, $B$ beats $C$ and $C$ beats $A$. Asymmetric games, on the other hand, very readily give rise to indefinite cyclical behaviour (see Appendix J). Although it is hard to point to examples, perhaps because of the long time-scales involved, the prediction is so clear that it would be odd if examples are not found.

The second situation in which a game theory model obliges one to think of change rather than constancy is when, as is often the case, a game has more than one ESS. Then, in order to account for the present state of a population, one has to allow for initial conditions – that is, for the state of the ancestral population. This is particularly clear in the analysis of parental care (p. 126).

Evolution is a historical process; it is a unique sequence of events. This raises special difficulties in formulating and testing scientific theories, but I do not think the difficulties are insuperable. There are two kinds of theories which can be proposed: general theories which say something about the mechanisms underlying the whole process, and specific theories accounting for particular events. Examples of general theories are 'all previous history is the history of class struggle', and 'evolution is the result of the natural selection of variations which in their origin are non-adaptive'. Evolutionary game theory is not of this kind. It assumes that evolutionary change is caused by natural selection within populations. Rather, game theory is an aid to formulating theories of the second kind; that is, theories to account for particular evolutionary events. More precisely, it is concerned with theories which claim to identify the selective forces responsible for the evolution of particular traits or groups of traits.

It has sometimes been argued that theories of this second, specific, kind are untestable, because it is impossible to run the historical process again with some one factor changed, to see whether the result is different. This misses the point that any causal explanation makes assumptions which can be tested. For example, in his *The Revolt of the Netherlands*, Geyl (1949) discusses why it was that the northern part of the Netherlands achieved independence when the south did not. The most commonly held explanation had been that the population of the north were mainly Protestant and of the south Catholic. Geyl shows that this explanation is wrong, because at the outbreak of the revolt, the proportion of Catholics did not differ between the two regions. Hypotheses about the causes of particular evolutionary events are likewise falsifiable. For example, the hypothesis that size dimorphism in the primates evolved because it reduces ecological competition between mates is almost certainly false, because dimorphism is large in polygynous and promiscuous mammals and absent in monogamous ones (Clutton-Brock, Harvey & Rudder, 1977); the hypothesis may be correct for some bird groups (Selander, 1972).

I think it would be a mistake, however, to stick too rigidly to the criterion of falsifiability when judging theories in population biology. For example, Volterra's equations for the dynamics of a predator and prey species are hardly falsifiable. In a sense they are manifestly false, since they make no allowance for age structure, for spatial distribution, or for many other necessary features of real situations. Their merit is to show that even the simplest possible model of such an interaction leads to sustained oscillation – a conclusion it would have been hard to reach by purely verbal reasoning. If, however, one were to apply this idea in a particular case, and propose, for example, that the oscillations in numbers of Canadian fur-bearing mammals is driven by the interactions between hare and lynx, that would be an empirically falsifiable hypothesis.

Thus there is a contrast between simple models, which are not testable but which may be of heuristic value, and applications of those models to the real world, when testability is an essential requirement.

# 2 *The basic model*

This chapter aims to make clear the assumptions lying behind evolutionary game theory. I will be surprised if it is fully successful. When I first wrote on the applications of game theory to evolution (Maynard Smith & Price, 1973), I was unaware of many of the assumptions being made and of many of the distinctions between different kinds of games which ought to be drawn. No doubt many confusions and obscurities remain, but at least they are fewer than they were.

In this chapter, I introduce the concept of an 'evolutionarily stable strategy', or ESS. A 'strategy' is a behavioural phenotype; i.e. it is a specification of what an individual will do in any situation in which it may find itself. An ESS is a strategy such that, if all the members of a population adopt it, then no mutant strategy could invade the population under the influence of natural selection. The concept is couched in terms of a 'strategy' because it arose in the context of animal behaviour. The idea, however, can be applied equally well to any kind of phenotypic variation, and the word strategy could be replaced by the word phenotype; for example, a strategy could be the growth form of a plant, or the age at first reproduction, or the relative numbers of sons and daughters produced by a parent.

The definition of an ESS as an uninvadable strategy can be made more precise in particular cases; that is, if precise assumptions are made about the nature of the evolving population. Section A of this chapter describes the context in which an ESS was first defined by Maynard Smith & Price (1973), and leads to the mathematical conditions (2.4*a*, *b*) for uninvadability. The essential features of this model are that the population is infinite, that reproduction is asexual, and that pairwise contests take place between two opponents, which do not differ in any way discernible to themselves before the contest starts (i.e. 'symmetric' contests). It is also assumed that there is a finite set of alternative strategies, so that the game can be expressed in matrix form; this assumption will be relaxed in Chapter 3.

Still using this model of pairwise contests, I then contrast the concept of an ESS with that of a population in an evolutionarily stable state. The distinction is as follows. Suppose that the stable strategy for some particular game requires an individual to do sometimes one thing and sometimes another – e.g. to do $I$ with probability $P$, and $J$ with probability $1-P$. An individual with a variable behaviour of this kind is said to adopt a mixed strategy, and the uninvadable strategy is a mixed ESS. Alternatively, a population might consist of some individuals which always do $A$ and others which always do $B$. Such a population might evolve to a stable equilibrium with both types present – that is, to an evolutionarily stable polymorphic state. The question then arises whether the probabilities in the two cases correspond; that is, if the mixed ESS is to do $I$ with probability $P$, is it also true that a stable polymorphic population contains a proportion $P$ of individuals which always do $I$? This question is discussed in section A below, and in Appendix D, for the case of asexual (or one-locus haploid) inheritance; the more difficult but realistic case of sexual diploids is postponed to Chapter 4.

Section B reviews the assumptions made in the model, and indicates how they might be relaxed or broadened. Section C considers a particular extension of the model, in which an individual is 'playing the field'; that is, its success depends, not on a contest with a single opponent, but on the aggregate behaviour of other members of the population as a whole, or some section of it. This is the appropriate extension of the model for such applications as the evolution of the sex ratio, of dispersal, of life history strategies, or of plant growth. The conditions for a strategy to be an ESS for this extended model are given in equations (2.9).

### A The Hawk–Dove game

Imagine that two animals are contesting a resource of value $V$. By 'value', I mean that the Darwinian fitness of an individual obtaining the resource would be increased by $V$. Note that the individual which does not obtain the resource need not have zero fitness. Imagine, for example, that the 'resource' is a territory in a favourable habitat, and that there is adequate space in a less favourable habitat in which losers can breed. Suppose, also, that animals with a territory in a favourable habitat produce, on average, 5 offspring, and that those breeding in the less favourable habitat produce 3 offspring. Then $V$ would equal $5-3=2$ offspring. Thus $V$ is the *gain* in fitness to the winner, and losers do not have zero fitness. During the contest an animal can behave in one of three ways, 'display', 'escalate' and 'retreat'. An animal which displays does not injure its opponent; one which escalates may succeed in doing so. An animal which retreats abandons the resource to its opponent.

In real contests, animals may switch from one behaviour to another in a complex manner. For the moment, however, I suppose that individuals in a given contest adopt one of two 'strategies'; for the time being, I assume that a particular individual always behaves in the same way.

'Hawk': escalate and continue until injured or until opponent retreats.

'Dove': display; retreat at once if opponent escalates. If two opponents both escalate, it is asumed that, sooner or later, one is injured and forced to retreat. Alternatively, one could suppose that both suffer some injury, but for the moment I am seeking the simplest possible model. Injury reduces fitness by a cost, $C$.

Table 1. *Payoffs for the Hawk–Dove game*

| | $H$ | $D$ |
|---|---|---|
| $H$ | $\frac{1}{2}(V-C)$ | $V$ |
| $D$ | 0 | $V/2$ |

Writing $H$ and $D$ for Hawk and Dove, it is now possible to write down the 'payoff matrix' shown in Table 1. In this matrix, the entries are the payoffs, or changes of fitness arising from the contest, to the individual adopting the strategy on the left, if his opponent adopts the strategy above. Some further assumptions were made in writing down the matrix, as follows:

(i) *Hawk v. Hawk* Each contestant has a 50% chance of injuring its opponent and obtaining the resource, $V$, and a 50% chance of being injured. Thus it has been assumed that the factors, genetic or otherwise, determining behaviour are independent of those which determine success or failure in an escalated contest. Later, in Chapter 8, I discuss contests in which differences, for example in size, which influence success in an escalated contest can be detected by the contestants.

(ii) *Hawk v. Dove* Hawk obtains the resource, and Dove retreats before being injured. Note that the entry of zero for Dove does *not* mean that Doves, in a population of Hawks, have zero fitness: it means that the fitness of a Dove does not alter as a result of a contest with a Hawk.

In the imaginary example, described above, of a contest over a territory, the fitness of a Dove, after a contest with a Hawk, would be 3 offspring.

(iii) *Dove v. Dove* The resource is shared equally by the two contestants. If the resource is indivisible, the contestants might waste much time displaying; such contests are analysed in Chapter 3.

Now imagine an infinite population of individuals, each adopting the strategy $H$ or $D$, pairing off at random. Before the contest, all individuals have a fitness $W_0$.

Let $p$ = frequency of $H$ strategists in the population,

$W(H)$, $W(D)$ = fitness of $H$ and $D$ strategists respectively,

and $E(H,D)$ = payoff to individual adopting $H$ against a $D$ opponent (and a similar notation for other strategy pairs).

Then if each individual engages in one contest,

$$\left.\begin{aligned} W(H) &= W_0 + p\,E(H,H) + (1-p)\,E(H,D), \\ W(D) &= W_0 + p\,E(D,H) + (1-p)\,E(D,D). \end{aligned}\right\} \quad (2.1)$$

It is then supposed that individuals reproduce their kind asexually, in numbers proportional to their fitnesses. The frequency $p'$ of Hawks in the next generation is

$$p' = p\,W(H)/\bar{W}, \quad (2.2)$$

where

$$\bar{W} = p\,W(H) + (1-p)\,W(D).$$

Equation (2.2) describes the dynamics of the population. Knowing the values of $V$ and $C$, and the initial frequency of $H$, it would be a simple matter to calculate numerically how the population changes in time. It is more fruitful, however, to ask what are the stable states, if any, towards which the population will evolve. The stability criteria will first be derived for the general case, in which more than two strategies are possible, and then applied to the two-strategy Hawk–Dove game.

If $I$ is a stable strategy,* it must have the property that, if almost all members of the population adopt $I$, then the fitness of these typical members is greater than that of any possible mutant; otherwise, the mutant could invade the population, and $I$ would not be stable. Thus consider a population consisting mainly of $I$, with a small frequency $p$ of some mutant $J$. Then, as in (2.1),

$$\left.\begin{aligned} W(I) &= W_0+(1-p)\,E(I,I)+p\,E(I,J), \\ W(J) &= W_0+(1-p)\,E(J,I)+p\,E(J,J). \end{aligned}\right\} \qquad (2.3)$$

Since $I$ is stable, $W(I) > W(J)$. Since $p \ll 1$, this requires, for all $J \neq I$,

*either* $\quad E(I,I) > E(J,I) \qquad (2.4a)$

*or* $\quad E(I,I) = E(J,I)$ *and* $E(I,J) > E(J,J). \qquad (2.4b)$

These conditions were given by Maynard Smith & Price (1973).

Any strategy satisfying (2.4) is an 'evolutionarily stable strategy', or ESS, as defined at the beginning of this chapter. Conditions (2.4*a*, *b*) will be referred to as the 'standard conditions' for an ESS, but it should be clear that they apply only to the particular model just described, with an infinite population, asexual inheritance and pairwise contests.

We now use these conditions to find the ESS of the Hawk–Dove game.

Clearly, $D$ is not an ESS, because $E(D,D) < E(H,D)$; a population of Doves can be invaded by a Hawk mutant.

* The distinction between a stable strategy and a stable state of the population is discussed further on pp. 16–17 and Appendix D.

$H$ is an ESS if $\frac{1}{2}(V-C)>0$, or $V>C$. In other words, if it is worth risking injury to obtain the resource, $H$ is the only sensible strategy.

But what if $V<C$? Neither $H$ nor $D$ is an ESS. We can proceed in two ways. We could ask: what would happen to a population of Hawks and Doves? I shall return to this question later in this chapter, but first I want to ask what will happen if an individual can play sometimes $H$ and sometimes $D$. Thus let strategy $I$ be defined as 'play $H$ with probability $P$, and $D$ with probability $(1-P)$'; when an individual reproduces, it transmits to its offspring, not $H$ or $D$, but the probability $P$ of playing $H$. It does not matter whether each individual plays many games during its life, with probability $P$ of playing $H$ on each occasion, the payoffs from different games being additive, or whether each individual plays only one game, $P$ then being the probability that individuals of a particular genotype play $H$.

Such a strategy $I$, which chooses randomly from a set of possible actions, is called a 'mixed' strategy; this contrasts with a 'pure' strategy, such as Hawk, which contains no stochastic element.

Is there a value of $P$ such that $I$ is an ESS? To answer this question, we make use of a theorem proved by Bishop & Cannings (1978), which states:

If $I$ is a mixed ESS which includes, with non-zero probability, the pure strategies $A,B,C,\ldots$, then

$$E(A,I) = E(B,I) = E(C,I)\ldots = E(I,I).$$

The reason for this can be seen intuitively as follows. If $E(A,I)>E(B,I)$ then surely it would pay to adopt $A$ more often and $B$ less often. If so, then $I$ would not be an ESS. Hence, if $I$ is an ESS, the expected payoffs to the various strategies composing $I$ must be equal. A more precise formulation and proof of the theorem is given in Appendix C. Its importance in the present context is that, if there *is* a value $P$ which makes $I$ an ESS of the Hawk–Dove game, we can find it by solving the equation

$$E(H,I) = E(D,I),$$

therefore

$$P\,E(H,H)+(1-P)\,E(H,D) = PE(D,H)+(1-P)\,E(D,D), \tag{2.5}$$

therefore

$$\tfrac{1}{2}(V-C)P+V(1-P) = \tfrac{1}{2}V(1-P),$$

or

$$P = V/C. \qquad (2.6)$$

More generally, for the matrix:

|  | $I$ | $J$ |
|---|---|---|
| $I$ | $a$ | $b$ |
| $J$ | $c$ | $d$, |

there is a mixed ESS if $a<c$ and $d<b$, the ESS being to adopt $I$ with probability

$$P = \frac{(b-d)}{(b+c-a-d)}. \qquad (2.7)$$

If there is an ESS of the form $I = PH+(1-P)D$, then $P$ is given by equation (2.6). We still have to prove, however, that $I$ satisfies equations (2.4*b*). Thus $E(H,I) = E(D,I) = E(I,I)$, and therefore stability requires that $E(I,D) > E(D,D)$ and $E(I,H) > E(H,H)$. To check this:

$$E(I,D) = PV+\tfrac{1}{2}(1-P)V > E(D,D).$$

and

$$E(I,H) = \tfrac{1}{2}P(V-C) > E(H,H), \text{ since } V < C.$$

Thus we have shown that, when $V<C$, a mixed strategy with $P=V/C$ is evolutionarily stable. The first conclusion from our model, then, is that, in contests in which the cost of injury is high relative to the rewards of victory, we expect to find mixed strategies. The model is so oversimplified that the conclusion must be treated with reserve. Field data bearing on it are discussed in Chapters 6 and 7, after some possible complications have been analysed theoretically.

The attainment of a mixed ESS depends on the assumption that a genotype can exist which specifies the mixed strategy and which can breed true. I now return to the question: what would happen to a population of pure Hawks and pure Doves? We have already seen that, if $V<C$, there can be no pure ESS. There may, however, be a stable genetic polymorphism; i.e. there may be a mixture of pure-breeding Hawks and Doves which is genetically stable.

Consider, then, a population consisting of $H$ and $D$ in frequencies $p$

and $1-p$. At equilibrium, the fitnesses $W(H)$ and $W(D)$ must be equal. That is

$$pE(H,H)+(1-p)E(H,D) = pE(D,H)+(1-p)E(D,D). \quad (2.8)$$

Equation (2.8) is identical to equation (2.5), with $p$ replacing $P$. Thus if $P$ gives the frequency of $H$ in a mixed ESS, and $p$ the frequency of $H$ in a population at genetic equilibrium, then $p=P$. This conclusion holds also if there are more than two pure strategies. But is the genetic polymorphism stable? When there are only two pure strategies, if the mixed strategy is stable then so is the genetic polymorphism; thus, for the Hawk–Dove game, a genetic polymorphism with a frequency of $p=V/C$ of pure Hawk is stable.

Unhappily, if there are more than two pure strategies, this simple conclusion no longer holds. It is possible for a mixed ESS to be stable but the corresponding polymorphism to be unstable, and vice versa. The problem of stability is discussed further in Appendix D; it is mainly of mathematical interest, if only because the stability of a polymorphism in an asexual population is a problem different from that of the stability of a sexual diploid population (see Chapter 4, section A).

I want now to extend the Hawk–Dove game by including more complex strategies. It will be convenient to replace the algebraic payoffs $V$ and $C$ by numerical ones; since only inequalities matter in determining qualitative outcomes, this makes things easier to follow without losing anything. Taking $V=2$ and $C=4$, there is a mixed ESS with $P=\frac{1}{2}$; the payoff matrix is

| | $H$ | $D$ |
|---|---|---|
| $H$ | $-1$ | 2 |
| $D$ | 0 | 1. |

Suppose now that we introduce a third strategy, $R$ or 'Retaliator'. $R$ behaves like a Dove against another Dove, but, if its opponent escalates, $R$ escalates also and acts like a Hawk. The payoff matrix is shown in Table 2*a*.

This more general version of the Hawk–Dove game and, in particular, the stability of retaliation is treated in more detail in Appendix E, which, I hope, corrects some of the errors I have made in earlier discussions of this problem. The game is discussed here to

Table 2. *The Hawk–Dove–Retaliator game*

| | *a* | | | | *b* | | |
|---|---|---|---|---|---|---|---|
| | *H* | *D* | *R* | | *H* | *D* | *R* |
| *H* | −1 | 2 | −1 | *H* | −1 | 2 | −1 |
| *D* | 0 | 1 | 1 | *D* | 0 | 1 | 0.9 |
| *R* | −1 | 1 | 1 | *R* | −1 | 1.1 | 1 |

illustrate how games with more than two strategies can be analysed. The matrix in Table 2*a* is awkward to analyse because, in the absence of Hawk, *D* and *R* are identical. It is shown in Appendix E that the only ESS is the mixed strategy, $I=\frac{1}{2}H+\frac{1}{2}D$.

The payoff matrix in Table 2*b* may be more realistic; it assumes that, in a contest between *D* and *R*, the Retaliator does, at least occasionally, discover that its opponent is unwilling to escalate, and takes advantage of this, so that, in *D* v. *R* contests, *R* does a little better and *D* a little worse. It is easy to see that *R* is now an ESS, because $E(R,R)$ is greater than either $E(D,R)$ or $E(H,R)$. Hence neither *D* nor *H*, nor any mixture of the two, could invade an *R* population. In general, if any entry on the diagonal of a payoff matrix is greater than all other entries in the same column, then the corresponding pure strategy is an ESS.

But is there any other ESS? In particular, what of $I=\frac{1}{2}H+\frac{1}{2}D$? Following the usual rules:

$$E(H,I) = -\tfrac{1}{2}+\tfrac{1}{2}\cdot 2 = \tfrac{1}{2},$$
$$E(D,I) = \tfrac{1}{2}\cdot 0+\tfrac{1}{2} = \tfrac{1}{2},$$

and hence $E(I,I)=\frac{1}{2}$. Note that, as required of a mixed ESS, $E(H,I)=E(D,I)$.

$$E(R,I) = -\tfrac{1}{2}+1\cdot 1\times\tfrac{1}{2} = 0.05.$$

The matrix in Table 2*b*, then, has two ESS's, $I=\frac{1}{2}H+\frac{1}{2}D$ and *R*. A population could evolve to either, depending on its initial composition.

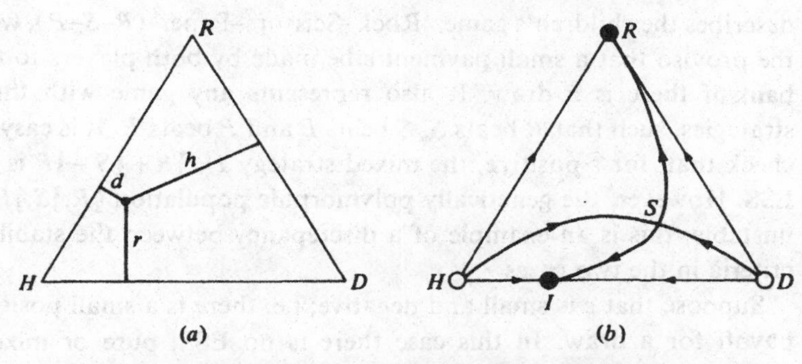


Figure 1. The Hawk–Dove–Retaliator game. (*a*) Representation of the state of a polymorphic population; *h*, *d* and *r* are the frequencies of pure *H*, *D* and *R* respectively. (*b*) Flows for the *H–D–R* game given in Table 2. There are attractors at *I* and *R* and a saddle point at *S*.

In picturing the dynamics of a game with three pure strategies, it is convenient to plot the state of the population as a point in an equilateral triangle, and then to plot the trajectories followed by the population, as in Figure 1. Of course, such a diagram can only represent the frequencies of the three pure strategies in a polymorphic population. In this case, however, there is a correspondence between the stable states of the polymorphic population and the stable strategies when mixed strategies are possible. Thus there are two stable states: pure *R*, and a polymorphism with equal frequencies of *H* and *D*, the latter corresponding to the mixed ESS, $I = \frac{1}{2}H + \frac{1}{2}D$.

A game with only two pure strategies always has at least one ESS (Appendix B); but if there are three or more pure strategies, there may be no ESS. Consider, for example, the matrix in Table 3. This

Table 3. *The Rock–Scissors–Paper game*

| | *R* | *S* | *P* |
|---|---|---|---|
| *R* | $-\varepsilon$ | 1 | $-1$ |
| *S* | $-1$ | $-\varepsilon$ | 1 |
| *P* | 1 | $-1$ | $-\varepsilon$ |

describes the children's game, 'Rock–Scissors–Paper' ($R$–$S$–$P$), with the proviso that a small payment $\varepsilon$ be made by both players to the bank if there is a draw. It also represents any game with three strategies, such that $R$ beats $S$, $S$ beats $P$ and $P$ beats $R$. It is easy to check that, for $\varepsilon$ positive, the mixed strategy $I=\frac{1}{3}R+\frac{1}{3}S+\frac{1}{3}P$ is an ESS. However, the genetically polymorphic population $\frac{1}{3}R, \frac{1}{3}S, \frac{1}{3}P$ is unstable; this is an example of a discrepancy between the stability criteria in the two cases.

Suppose that $\varepsilon$ is small and negative; i.e. there is a small positive payoff for a draw. In this case there is no ESS, pure or mixed. In the absence of an ESS, the population will cycle indefinitely, $P \rightarrow S \rightarrow R \rightarrow P \rightarrow \ldots$. I cannot decide whether there are intraspecific contest situations likely to lead to such indefinite cycles; comparable cycles, in asymmetric games, are discussed on p. 130 and Appendix J.

## B A review of the assumptions

### *An infinite random-mixing population*

If, as will commonly be the case, individuals do not move far from where they were born, this will alter the model in various ways.

First, opponents will have some degree of genetic relatedness. An analysis of games between relatives is given in Appendix F. The problem turns out to be far from straightforward. At a qualitative level, however, the conclusion is the commonsense one, that animals will behave in a more Dove-like and less Hawk-like manner.

Secondly, an individual may have a succession of contests against the same opponent. If there is no learning from experience, this will not alter the conclusions. If there is learning, then the 'strategies' which have to be considered when seeking an ESS are no longer fixed behaviour patterns, but 'learning rules'; the evolution of learning rules is discussed in Chapter 5.

Thirdly, it is possible that the population whose evolution is being considered is not only finite but small. If so, the basic model must be altered, because mutants cannot be very rare. Finite population games have been considered by Riley (1978).

### *Asexual reproduction*

Most species whose behaviour is of interest are sexual and diploid, whereas the model outlined above assumes asexual reproduction. This discrepancy is unlikely to matter in practice. When reasoning about the function of some behavioural trait, some assumption must be made about the range of phenotypes possible to the species; i.e. the 'strategy set'. This may be based in part on knowledge of the range of actual variability in the species or in related species and in part on guesswork or common sense. It is most unlikely to be based on a knowledge of the genetic basis of the behavioural variability. Therefore a simple assumption of 'like begets like' is often more sensible than a detailed assumption about the genetic basis. A case where there seems no escape from detailed genetic hypotheses is discussed in Chapter 10, section D.

It is, however, important to be able to show, for simple model situations, that the results of parthenogenesis and of diploid inheritance are similar. This is done for a particular case by Maynard Smith (1981), and in Chapter 4, section A. Briefly, an infinite random-mating diploid population plays a game with two pure strategies; $P^*$ represents the frequency of one strategy at the ESS (i.e. $P^*$ is given by equation 2.7). The actual frequency with which an individual adopts that strategy is determined by two alleles, *A* and *a*, being $P_0$, $P_1$ and $P_2$ in *AA*, *Aa* and *aa*, respectively.

If $P_0 \geqslant P_1 \geqslant P_2$ (i.e. no overdominance), then the population will evolve to the ESS provided $P^*$ lies between $P_0$ and $P_2$. If $P^*$ lies outside that range, then obviously the population cannot evolve to $P^*$, but it will become fixed for the homozygote lying closest to the ESS. If there is overdominance, things are more complex, but it is still true that the population will usually evolve to an ESS if the genetic system permits, and otherwise approaches it as closely as it can. Eshel (1981*b*) has shown that a diploid population will evolve to the ESS for a wide range of genetic structures, although it is not true for the most general ones.

In general, as the number of loci, or number of alleles per locus, increases, it becomes more likely that a population will reach an ESS (Slatkin, 1979). If the ESS requires a range of phenotypes, achievable only by a genetic polymorphism and not by a mixed strategy, then the genetic system may prevent the phenotypes existing in the appropriate frequencies. As an example, the ESS for the 'war of attrition' discussed in the next chapter requires a phenotypic distribution which could not easily be generated by a polygenic system. Of course, no difficulty arises if an individual can adopt a mixed strategy.

### *Symmetric and asymmetric contests*

The Hawk–Dove game analysed above is symmetrical. That is to say, the two contestants start in identical situations: they have the same choice of strategies and the same prospective payoffs. There may be a difference in size or strength between them, which would influence the outcome of an escalated contest, but if so it is not known to the contestants and therefore cannot affect their choice of strategies.

Most actual contests, however, are asymmetric. They may be between a male and a female, between an old and young, or a small and large individual, or between the owner of a resource and a non-owner. An asymmetry may be perceived beforehand by the contestants; if so, it can and usually will influence the choice of action. This is most obviously so if the asymmetry alters the payoffs, or affects the likely outcome of an escalated contest. It is equally true, although less obvious, that an asymmetry which does not alter either payoffs or success in escalation can determine the choice of action.

Table 4. *The Hawk–Dove–Bourgeois game*

| | *H* | *D* | *B* |
|---|---|---|---|
| *H* | −1 | 2 | 0.5 |
| *D* | 0 | 1 | 0.5 |
| *B* | −0.5 | 1.5 | 1.0 |

Thus, consider a contest between the owner of a territory and an intruder. In practice, the value of the territory may be greater to the owner because of learnt local knowledge, and it is also possible that ownership confers an advantage in an escalated contest. For simplicity, however, I shall ignore these effects. Let us introduce into the Hawk–Dove game a third strategy, *B* or 'Bourgeois'; i.e. 'if owner, play Hawk; if intruder, play Dove'. The payoff matrix is shown in Table 4.

Note that it is always the case, when two *B* strategists meet, that one is the owner and the other intruder. I have assumed in filling in the matrix that each strategy type is owner and intruder equally frequently. That is, genes determining behaviour are independent of the factors, genetic or environmental, determining ownership.

It is clear that $B$ is an ESS, and easy to check that it is the only ESS of this game. Thus an asymmetry of ownership will be used as a conventional one to settle the contest, even when ownership alters neither the payoffs nor success in fighting. The same is true of any other asymmetry, provided it is unambiguously perceived by both contestants. Asymmetric contests are discussed in detail in Chapters 8–10.

*Pairwise contests*

The Hawk–Dove model, and more complex models expressed in payoff matrix form, suppose that an individual engages in one or more pairwise contests; if more than one contest occurs, payoffs are assumed to be combined additively. Such a model can be applied to agonistic encounters between pairs, or, in asymmetric form, to contests between mates or between parent and offspring. There are many situations, however, in which an individual is, in effect, competing not against an individual opponent but against the population as a whole, or some section of it. Such cases can loosely be described as 'playing the field'. Examples include the evolution of the sex ratio (Fisher, 1930; Shaw & Mohler, 1953; Hamilton, 1967), of dispersal (Fretwell, 1972; Hamilton & May, 1977), of competition between plants (since each plant competes against all its neighbours, not against a single opponent), and many other examples. In fact, such contests against the field are probably more widespread and important than pairwise contests; it therefore seems appropriate to discuss them under a separate head.

## C An extended model – playing the field

We can extend the concept of an 'unbeatable strategy' (Hamilton, 1967) or an 'evolutionarily stable strategy', to cases in which the payoff to an individual adopting a particular strategy depends, not on the strategy adopted by one or a series of individual opponents, but on some average property of the population as a whole, or some section of it.

Table 5. *Fitness matrix for the extended model*

| | | Population | |
|---|---|---|---|
| | | $I$ | $J$ |
| Mutant | $I$ | $W(I,I)$ | $W(I,J)$ |
| | $J$ | $W(J,I)$ | $W(J,J)$ |

How should an ESS be defined when individuals are playing the field? This question has been treated by P. Hammerstein (personal communication), and I have followed his proposal. Let the fitness of a single $A$ strategist in a population of $B$ strategists be written $W(A,B)$. Clearly, $I$ will be an ESS if, for all $J \neq I$, $W(J,I) < W(I,I)$. But what if $W(J,I) = W(I,I)$? We then need that $W(J) < W(I)$ in a population of $I$ strategists containing a small proportion $q$ of $J$ strategists. We define $W(J,P_{q,J,I})$ as the fitness of a $J$ strategist in a population $P$ consisting of $qJ+(1-q)I$. The conditions for $I$ to be an ESS then are, for all $J \neq I$,

$$\left.\begin{array}{ll} \textit{either} & W(J,I) < W(I,I) \\ \textit{or} & W(J,I) = W(I,I) \\ \text{and, for small } q, & \\ & W(J,P_{q,J,I}) < W(I,P_{q,J,I}). \end{array}\right\} \quad (2.9)$$

If only two strategies are possible, $I$ and $J$, we can draw up the fitness matrix in Table 5.

If $W(J,I) < W(I,I)$, then $I$ is an ESS; if $W(I,J) < W(J,J)$, then $J$ is an ESS. If neither of these inequalities hold, then the ESS is a mixture of $I$ and $J$. It would be wrong though, to think that the proportions of the two strategies at the ESS are necessarily given by equation (2.7). This would be true only if the fitness of an individual $I$ in a population consisting of a mixture $I$ and $J$ in proportion $P$ to $1-P$ were given by the linear sum $PW(I,I)+(1-P)W(I,J)$, and this is not necessarily so.

These points can best be illustrated by considering the simplest form of the sex ratio game, in which a female can produce a total of $N$

Table 6. *Fitness matrix for the sex ratio game*

| | | Population | |
|---|---|---|---|
| | | $s_1 = 0.1$ | $s_2 = 0.6$ |
| Mutant | $s_1 = 0.1$ | 1.8 | 0.967 |
| | $s_2 = 0.6$ | 5.8 | 0.8 |

offspring, in the ratio $s$ males to $(1-s)$ females. If we measure 'fitness' as expected number of grandchildren, then in a random-mating population of sex ratio $s'$, we have

$$W(s,s') = N^2\left[1-s+s\frac{(1-s')}{s'}\right],$$

and $$W(s',s') = 2N^2(1-s'). \tag{2.10}$$

If we then consider a population containing two types of female, producing sex ratios $s_1 = 0.1$ and $s_2 = 0.6$, we have the fitness matrix in Table 6.

It is apparent that neither $s_1$ nor $s_2$ is an ESS. If, without justification, we were to calculate $P$ from equation (2.7), we would conclude, wrongly, that the stable state consisted of 1/25 of $s_1$ and 24/25 of $s_2$, giving a population sex ratio of $14.5/25 = 0.58$. In fact, the stable population sex ratio is 0.5.

Supposing that only these two kinds of females existed, the correct way to find the ESS is as follows. Let $\hat{s}$ be the population sex ratio at equilibrium.

Then $W(s_1,\hat{s}) = W(s_2,\hat{s})$, or

$$1-0.1+0.1(1-\hat{s})/\hat{s} = 1-0.6+0.6(1-\hat{s})/\hat{s},$$

or $\hat{s} = 0.5$,

requiring

$$0.2s_1 + 0.8s_2.$$

More generally, suppose individual females can produce any sex ratio between 0 and 1. We seek a sex ratio $s^*$, which is an ESS in the sense of being uninvadable by any mutant with $s \neq s^*$. That is, $W(s^*, s^*) > W(s,s^*)$ for $s \neq s^*$. Provided that $W$ is differentiable, we can find $s^*$ from the condition

$$[\partial W(s,s^*)/\partial s]_{s = s^*} = 0. \tag{2.11}$$

Applying this condition to equation (2.10) gives $s^* = 0.5$, as expected. We can use equations (2.9) to check the stability of $s^* = 0.5$, as follows:

Let $\quad s' = qs + (1-q)s^*$, where $s \neq s^*$.

Then from equation (2.10),

$$W(s,s') = N^2\left[1 - s + s\frac{(1-s')}{s'}\right],$$

and

$$W(s^*,s') = N^2\left[1 - s^* + s^*\frac{(1-s')}{s'}\right].$$

It is then easy to show that, for $s \neq s^*$, the inequality $W(s,s') < W(s^*,s')$ holds.

To summarise the extended model, a strategy $I$ is an ESS provided that equations (2.9) are satisfied. If, in a game with two pure strategies, $I$ and $J$, neither satisfies equations (2.9), the ESS will be a mixed strategy; however, the relative frequencies of $I$ and $J$ at the equilibrium cannot be found from equation (2.7), but must be calculated from the equation $W(I, Pop) = W(J, Pop)$, where $Pop$ refers to the equilibrium population. If the strategy set is a continuous variable (e.g. the sex ratio, varying continuously from 0 to 1), the ESS can be found from a condition similar to equation (2.11); its stability must be checked by taking the second derivative, or in some other way.

The crucial step in analysing cases in which an individual is playing the field is to write down expressions corresponding to equation (2.10), giving the fitness of a rare mutant in a population of known composition. In the particular case of equation (2.10), the population is treated as infinite and without structure. This, however, is not a necessary restriction. For example, Hamilton (1967) sought the unbeatable sex ratio, $s^*$, when the offspring of $k$ females mate randomly *inter se*. The problem reduces to writing down an expression $W(s,s^*)$ for the fitness of an individual producing sex ratio $s$ when in a group with $k-1$ females producing a sex ratio $s^*$, and then applying condition (2.11). In other words, given that the other females in the group produce the sex ratio $s^*$, the best thing for the $k^{\text{th}}$ female is to do likewise.

To give another example of a structured population, consider competition between plants or sessile animals growing in a pure stand. We would seek a growth strategy $I$ such that, if all the neighbours of an individual were adopting $I$, the best strategy for the individual is also $I$. Mirmirani & Oster (1978) considered competition between annual plants which differed in the time at which they switched resources from growth to seed production. To find the evolutionarily stable time, $T^*$, it would be necessary to find $W(T,T^*)$, the seed production of an individual switching at time $T$ if surrounded by individuals switching at time $T^*$, and then to solve the equation $[\partial W(T,T^*)/\partial T]_* = 0$. Note that it would not be necessary to work out the fitness of individuals surrounded by a mixture of types.

As a summary of the ideas in this chapter, it might be helpful to read through the 'Explanation of main terms' on p. 204.

# 3 *The war of attrition*

In the last chapter, I assumed that two 'Doves' competing for a resource worth $V$ could share the resource. There will be many cases in which it will not be worth while to share a resource. For example, suppose two animals compete for a territory, and that there is no asymmetry, such as prior ownership, which can settle the matter.

Let $N$ = expected offspring to the owner of the territory,
$kN$ = expected offspring to the owner of half the territory ($k < 1$),
$n$ = expected offspring to an animal who does not compete but instead sets up a territory in a less satisfactory habitat ($n < N$).

If $n > kN$, it would not pay either contestant to share the territory. The payoff $V$ for obtaining the territory is $N-n$; note that it is *not* the expected fitness of an owner of the territory, but the *change* in fitness for winning.

Suppose, then, that $V = N-n$, and that the contest is settled without escalation. That is, the contestants display, and the owner is the one which persists longest. For how long should a contestant persist? If displaying cost nothing, the contestant should persist for ever, which is clearly absurd. In practice, to display must cost something, if only because to display for a long time is to delay the start of breeding.

I assume, therefore, that the cost of displaying increases with the length of the contest and is the same for the two contestants. The only choice open to an individual is to select a length of time for which he is prepared to continue, and an associated cost, $m$, he is prepared to pay. Thus if the two contestants, $A$ and $B$, select costs $m_A$ and $m_B$, respectively, the winner will be the one selecting the higher cost; however, he will not have to pay that cost, because the length of the contest is determined by the loser. Thus the payoffs are

| | Player *A* | Player *B* |
|---|---|---|
| $m_A > m_B$ | $V - m_B$ | $-m_B$ |
| $m_A = m_B$ | $(V/2) - m_B$ | $(V/2) - m_B$ |
| $m_A < m_B$ | $-m_A$ | $V - m_A$. |

This assumes that in the (infinitely unlikely) event that $m_A = m_B$, the contest is decided randomly. Given these payoffs, what choice of $m$ is evolutionarily stable?

Before answering this question, one biological point must be made. In assuming that the only possible choice of strategy is a choice of $m$, made before the contest, I have assumed that no relevant information (e.g. about what would happen in an escalated contest) is obtained during the contest. The problem of information transfer is crucial; it is discussed further on p. 35 and in later chapters.

Clearly, no pure strategy can be an ESS. Thus suppose the members of a population play $M$. Their average payoff is $(V/2) - M$. A mutant playing $M + \delta M$ would have an average payoff $V - M$, and could invade. If $M > (V/2)$, a mutant playing 0 could also invade.

Hence, if there is an ESS, it must be a mixed one. Let $I$ be a strategy defined by the probability density function $p(x)$. That is, the probability of accepting a cost between $x$ and $x + \delta x$ is $p(x)\delta x$. To find $p(x)$, we make use of the Bishop–Cannings theorem (Appendix C), which in the present context states that, if $m$ is a pure strategy in the 'support' of $I$ (i.e. $p(m) \neq 0$), then $E(m,I)$ is constant.

Now

$$E(m,I) = \int_0^m (V - x)p(x)\mathrm{d}x - \int_m^\infty m\, p(x)\mathrm{d}x.$$

We have to find $p(x)$ such that $\partial E(m,I)/\partial m = 0$, subject to the constraint $\int_0^\infty p(x)\mathrm{d}x = 1$. It is easy to confirm that

$$p(x) = (1/V)\mathrm{e}^{-x/V} \qquad (3.1)$$

is the required function. This shows that $I = p(x)$ is an equilibrium strategy; to show that it is stable, we must also show that

$$E(I,m) > E(m,m). \qquad (3.2)$$

This can easily be done if $m$ is a pure strategy (Maynard Smith, 1974); it has been proved by Bishop & Cannings (1978) for the case when $m$ can be any mixed strategy different from $I$.

The negative exponential form of equation (3.1) is intuitively appealing for the following reason. Since no information is exchanged, a contestant who has continued for time $t$, and whose opponent is still displaying, is in exactly the same state as far as *future* gains and losses are concerned as he was at time zero. Logically, therefore, he should make the same choice of future expenditure at time $t$ as at time zero; this requires a negative exponential distribution.

If cost is a linear function of time, then the times for which an individual is prepared to display will be distributed as a negative exponential. A stay time with this distribution, however, is not particularly strong evidence for a mixed ESS, since all that is needed to generate such a distribution is that the individual should have a constant probability of leaving per unit time. It must also be shown that the constant probability has the correct value – 'correct' here means the value which equalises the fitnesses of individuals with different stay times, as is true for the distribution given by equation (3.1). The work of Parker (1970*a*,*b*) and of Parker & Thompson (1980) on the dung fly *Scatophaga stercoraria* affords two examples, one of which may be a mixed ESS, and the other certainly is not.

Female dung flies come to fresh cowpats to lay their eggs. The males congregate at cowpats, and attempt to mate with arriving females. For how long should a male stay at a cowpat? Many females arrive at a fresh pat, and progressively fewer arrive as the pat grows stale. Therefore, a male which stays too long will meet few females. However, if most males stay only for a short time, a male which stays for a longer time will have a better chance of mating with those females which do arrive. Hence, if other males move it pays to stay, and vice versa. The contest is a frequency-dependent one similar, but not mathematically identical, to the war of attrition.

Parker (1970*a*) found that male stay times are exponentially distributed. Further, if female arrival rates are measured, it is found that male stay times are so distributed as to give the same expected success to males adopting different strategies (Figure 2). To get equal success rates, Parker had to suppose that the time required to find a

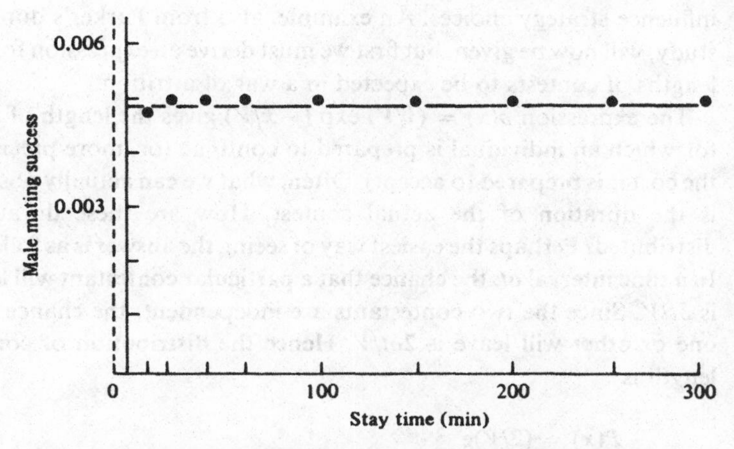


Figure 2. Estimated mating success of male dung flies, as a function of their stay times at cowpats, assuming it takes four minutes to find a fresh pat. (After Parker, 1970*a*.)

new pat, after leaving an old one, was four minutes. This was not an arbitrary choice made only to get a good fit; four minutes is the average time it takes males to arrive at a freshly deposited pat. Parker's data, therefore, provide a striking fit with the theory.

The mechanism by which this is achieved, however, is not known. There are at least three possibilities. First, the population may be genetically variable, with each male having a different genetically determined stay time. Secondly, all males may be alike, with an individually flexible stay time; since the distribution is, approximately, a negative exponential, all that this requires is that each male should have the same constant probability of leaving per unit time. Thirdly, and perhaps most plausible, males may adjust their stay times in the light of experience. It will be shown in Chapter 5 that learning can take a population to the ESS frequencies in a single generation, without genetic evolution. A learning mechanism would have the advantage of enabling males to adjust their behaviour as the density of cowpats changes.

It may not be accidental that Parker's data refer to a contest in which individuals are playing the field. In pairwise contests, asymmetries of size, ownership, sex, age etc. are likely to be perceived and to influence strategy choices. An example, also from Parker's dung fly study, will now be given, but first we must derive an expression for the lengths of contests to be expected in a war of attrition.

The expression $p(x) = (1/V) \exp(-x/V)$ gives the length of time for which an individual is prepared to continue (or, more precisely, the cost it is prepared to accept). Often, what we can actually observe is the duration of the actual contest. How are these durations distributed? Perhaps the easiest way of seeing the answer is as follows. In a time interval $\delta t$, the chance that a particular contestant will leave is $\delta t/V$. Since the two contestants are independent, the chance that one or other will leave is $2\delta t/V$. Hence the distribution of contest length is

$$P(x) = (2/V)e^{-2x/V}. \tag{3.3}$$

Thus contest lengths are also exponentially distributed, but with mean $V/2$ instead of $V$.

Parker & Thompson (1980) derived this result and applied it to a later stage of the contest between male dung flies. After mating, females stay on the dung laying eggs. The male remains on the back of the female during this period. In this way, he prevents a second male from copulating with the female; if a second copulation does occur, the second male's sperm fertilise 80% of the eggs laid subsequently.

While a female paired in this way is laying eggs, unmated males attempt to displace the male in possession (Parker, 1970*b*). Usually, an approaching male is deflected by the owner without a struggle. If, however, the approaching male manages to touch the female, a struggle often ensues, in which the intruder attempts to displace the owner. Parker & Thompson (1980) analyse these struggles. The durations are, approximately, exponentially distributed. Further, the relation between mean duration and estimated costs is at least consistent with a 'war of attrition' interpretation, although costs cannot be measured with any precision. The authors point out, though, that it would be quite wrong to interpret the contests in this way. Thus if, as in Figure 3, a distinction is made between those won by owners and by intruders, the contests are seen to be quite different, and the latter are far from exponential in distribution. Yet in the symmetric war of attrition the two distributions should be the same,

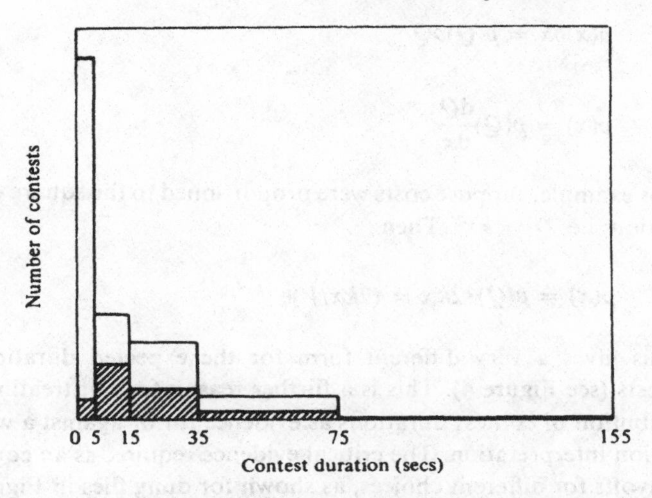


Figure 3. Observed lengths of contests between male dung flies. Open histogram, all data; cross-hatched histogram, contests in which the attacking male won. (After Parker & Thompson, 1980.)

and both exponential. The contest, clearly, is an asymmetric one, and should be analysed as such; this will be done on p. 121. It has been mentioned here as a warning; an exponential distribution of contest durations is an insufficient reason for regarding a contest as a symmetric war of attrition.

It may be that cost is not a linear function of time. If so, the contest can still be analysed in the same way, but its duration will no longer be exponentially distributed (Norman, Taylor & Robertson, 1977). Thus suppose the cost $Q$ is some function $q(x)$ of the time $x$ for which the contest lasts. The contestants can be thought of as choosing an acceptable cost, and, by exactly the same argument as that leading to equation (3.1), the stable distribution of choices will be

$$p(Q) = (1/V)e^{-Q/V}. \qquad (3.4)$$

What, then, will be the distribution of $x$? The probability that an individual will select a time between $x$ and $\delta x$ is the same as the probability that it will accept a cost between $q(x)$ and $q(x+\delta x)$. That is

$$p(x)\delta x = p(Q)\delta Q$$

or

$$p(x) = p(Q)\frac{dQ}{dx}.$$

For example, suppose costs were proportioned to the square of the duration; i.e. $Q = kx^2$. Then

$$p(x) = p(Q)\cdot 2kx = (2kx/V)e^{-kx^2/V}$$

This gives a very different form for the expected duration of contests (see Figure 4). This is a further reason for not treating the distribution of contest durations as evidence for or against a war of attrition interpretation. The critical evidence required is an equality of payoffs for different choices, as shown for dung flies in Figure 2.

Bishop & Cannings (1978) point out that the war of attrition model can be applied in a wide range of contexts, provided that:

(i) No relevant information is received during the contest, so that an action (i.e. a persistence time) can in effect be made at the start.

(ii) The winner is the contestant prepared to accept the higher cost.

(iii) The actual cost to both contestants is equal to the cost acceptable to the loser.

(iv) The range of possible actions must be a continuous one; the significance of this is discussed further on p. 105.

For example, cost might be measured by injury received during the contest. Such injury might be proportional to the length of the contest; alternatively, the strategy choice might be of a level to which the contestant would escalate, the amount of injury increasing as the level was raised. It need not even be the case that actual injury received is a function of duration or level of escalation, provided that the *risk* of injury (i.e. the 'expected' injury) is such a function. It is, however, a necessary feature of the model that injury should not be so great as to prevent a contestant from continuing. A crucial difference between the war of attrition and the Hawk–Dove game is that, in the former, an animal can almost guarantee victory by choosing a sufficiently high risk (although, of course, it cannot guarantee a positive payoff), whereas a Hawk meeting another Hawk has only an even chance of victory.

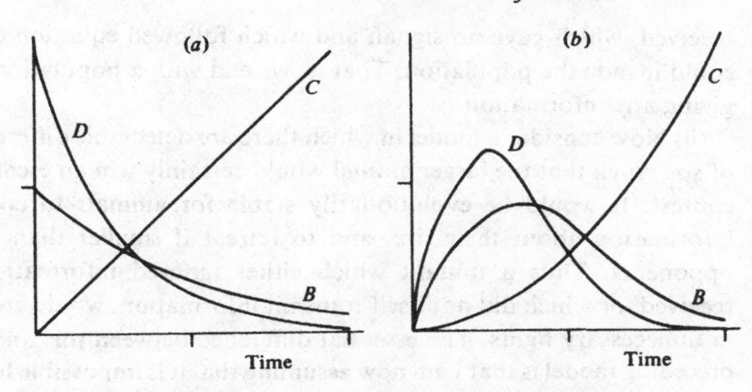


Figure 4. The war of attrition. Distribution of acceptable durations ($B$), and durations of contests ($D$), when ($a$) cost, $C$, is proportional to time, and ($b$) cost is proportional to the square of the time.

When discussing the persistence times of male dung flies at cowpats, the point was made that this is a contest in which each individual is playing the field, and that the reasonable fit with the war of attrition model is probably dependent on this fact; in pairwise contest, information transfer is likely to influence behaviour. The time has now come to discuss information transfer. It is convenient to start by considering two extreme models.

(i) There are no differences in size or weapons which can be detected by the contestants. There are, however, differences in motivation, leading contestant $A$ to choose cost $m_A$ and $B$ to choose cost $m_B$, where $m_A > m_B$, say. Would it not pay them both to signal the level they have chosen, and for $B$ then to retreat at once? Indeed, both would be better off; $A$ would gain $V$ instead of $V - m_B$, and $B$ would gain 0 instead of $-m_B$. Unfortunately, this signalling behaviour is not proof against 'lying'. Thus suppose we start with a population of individuals which select a value of $m$ according to equation (3.1), signal it accurately, and retreat at once if their opponent signals a higher value. A mutant which signals a large value $M$, but retreats if its opponent does not retreat at once, can invade such a population. Soon the population would consist of individuals signalling high values of $M$ which did not correspond to their actual future behaviour. At this stage, a mutant which ignored the signal it received, which gave no signal, and which followed equation (3.1), could invade the population. That is, we end with a population not giving any information.

(ii) Now consider a model in which there are detectable differences of size, such that the larger animal would certainly win an escalated contest. It would be evolutionarily stable for animals to convey information about their size, and to retreat if smaller than their opponents. Thus a mutant which either ignored information it received, or which did not itself transmit information, would engage in unnecessary fights. The essential difference between this and the preceding model is that I am now assuming that it is impossible for an animal to give false information about its size.

This distinction is crucial to an understanding of animal contests in general, and information transfer in particular. In the first model, it is possible for an animal to transmit any signal, at little or no cost, except in so far as there might be a cost exacted in the subsequent course of the contest. In the second model, it is impossible for an animal to transmit false information about its size, although there will certainly be selection for animals to appear as large as possible. Also, since larger animals win contests, there will be strong selection for increased size. There are also likely to be counteracting disadvantages to large size. An analysis of this situation is given in Chapter 11.

The problems of information transfer are discussed further in Chapters 9 and 12. For the present, the essential point is to distinguish two cases:

(i) Information about 'motivation' or 'intentions'. Because any message about motivation can be sent, with little cost, there is no reason why such messages should be accurate, and therefore no advantage in paying attention to them.

(ii) Information about 'Resource-Holding Power', or RHP (Parker, 1974*b*); RHP is a measure of the size, strength, weapons etc. which would enable an animal to win an escalated contest. It can be evolutionarily stable to transmit information about RHP, and to accept such information to settle a contest, provided two things are true. It must be impossible to transmit false information about RHP, and it must be expensive to acquire high RHP in the first place.

I turn now from the problem of information transfer to discuss cases in which the value of winning is not the same for the two

Table 7. *Supposed breeding success, in offspring successfully reared, of young and old birds*

| | Favourable habitat | Unfavourable habitat |
|---|---|---|
| First-year bird | 2 | 1 |
| Old bird | 4 | 2 |

contestants. An example is a contest for food between a hungry and a well-fed animal. To take a more complex example, suppose that two kinds of birds compete for a territory in a favourable habitat, and that the loser can establish a territory in a less favourable habitat without further contests. The two birds may be of different ages: for example, a first-year bird and an older bird. Suppose that expected breeding success is as shown in Table 7. The payoff for winning is then 2 to the older bird and 1 for the first-year bird.

Suppose first that the difference between young and old birds can be recognised unambiguously. Then, as Hammerstein (1981) pointed out, the contest should be analysed as three separate games: young v. young, old v. old, and young v. old. In the first two games, there are no payoff differences to worry about. The third game is a typical asymmetric game of the type discussed in Chapters 8 to 10; almost certainly, the age difference would be used as a cue to settle the contest.

Suppose, however, that the age of an opponent cannot be detected, so that a bird's behaviour can be influenced by its own age status, but not by its opponent's. This is an example of a game of imperfect information (discussed further in Chapter 12); each contestant has some information not available to its opponent. The earlier example of a contest between a hungry and a well-fed bird would be logically similar if a bird knew only its own state of hunger.

The problem of the war of attrition in which an individual knows the value of the resource to itself, but knows only the probability distribution of the value to its opponent, is analysed in Appendix G. Applied to the example of a territorial contest, the conclusions are as follows. Younger birds will select an acceptable cost, $m$, from a probability distribution ranging from zero to some threshold value, $T$, and older birds from a distribution ranging from $T$ to $\infty$. Thus, old birds will always win against young ones, but symmetric contests will be settled, as in a typical war of attrition, by the chance selection of a value of $m$ from the same distribution.

If there are only two categories of individual, there is a single threshold value $T$. If there are $N$ categories, for which the values of winning are $V_1 < V_2 < \ldots < V_N$, there will be $N$ non-overlapping probability distributions separated by $N-1$ threshold values. Contests will be won by the animal with the larger value.

What is the average payoff per contest in the war of attrition? For the simple case, with an ESS given by equation (3.1), it is easy to see that the average payoff is zero. Thus the defining characteristic of equation (3.1) is that the payoff for all values of $x$ is the same. This includes the payoff for $x = 0$, which is clearly zero. In other words, the average cost of a contest is equal to $V/2$, the average gain. This may at first sight seem an odd result. It does *not* mean, however, that animals have, on average, zero fitness. Thus suppose, for example, that all territorial contests were symmetric ones between older birds. The value of winning is 2 offspring, so the average cost will be 1 offspring, compared to an average breeding success of 3 in the favourable habitat and 1 in the unfavourable one.

Things are different, however, if the rewards are variable. It is still true that the average payoff is zero for that category with the lowest value for winning; it is positive for all other categories.

The essential feature of the 'variable rewards' model is that animals know the value of the resource to themselves, but not to their opponent. There is one rather strange example which may illustrate this model. This concerns the digger wasp *Sphex ichneumoneus*. Females of this species dig holes, which they then provision with katydids, before laying a single egg and sealing the burrow. Sometimes, instead of digging a burrow, a female will enter a burrow already dug by another wasp. The choice between these strategies is analysed (pp. 74–5) as an example of a mixed ESS. For the present, however, I want to concentrate on the fights which occur if two wasps who are provisioning the same burrow actually meet. Dawkins & Brockmann (1980) analyse Brockmann's data on 23 such fights observed in the field.

For each fight, it is known how long it lasted, who won, which wasp dug the hole, which was larger, and how many katydids each had supplied. Surprisingly, there was no significant advantage for the larger of the two, nor for the owner over the joiner. Eleven wasps fought more than once; there was no significant tendency for some wasps to be winners and others losers. What then does determine the outcome of fights? The hypothesis which best fits the facts is that a wasp fights for a length of time which increases with the number of katydids it has brought to the nest, and hence that the winner is the wasp which has brought most katydids. There is, of course, a correlation between the number of katydids brought by the loser and the total number present, but analysis shows that it is the number brought by the loser which is relevant in determining the length of a fight.

These results are what would be expected from a war of attrition with random rewards, provided that we assume that a wasp knows how many katydids it has supplied (presumably, by monitoring its own activity) but not the total number present. If this is so, then the value of a burrow is indeed an increasing function of what the individual has supplied, and the length of time the individual will fight should likewise increase with the number of katydids supplied.

# 4 *Games with genetic models*

The genetic assumption which underlies the standard ESS conditions of equation (2.4*a*,*b*), or (2.9) for the extended model, is that of parthenogenetic inheritance. Most populations of interest have sexual diploid inheritance. In many cases this does not matter. If the phenotype, pure or mixed, which satisfies the standard conditions is one which can be produced by a genetic homozygote, then a sexual population with that genotype will be stable against invasion by any mutant. Suppose, however, the ESS cannot be produced by a genetic homozygote. The question then arises of whether a genetically polymorphic population can generate the appropriate strategies in the ESS proportions, and if so whether such a population will be stable. Clearly the answer cannot always be yes. For example, suppose the three pure strategies $I$, $J$ and $K$ must be present in equal frequencies at an ESS. If they were produced, respectively, by the three genotypes $AA$, $Aa$ and $aa$ at a locus, then, in a random-mating population, there is no way this could happen. This problem is discussed, for a simple genetic model, in section A. It is treated more generally by Eshel (1981*b*).

An explicit genetic model may also be needed when the phenotypic trait of interest is itself concerned with the process of sexual reproduction, so that appropriate fitnesses can only be calculated for a sexual model. Examples of such traits are discussed in sections B and C. A much more general treatment of sexual allocation problems from a game theoretic point of view is given by Charnov (1979, 1981).

## A The two-strategy game with diploid inheritance

Imagine a game in which only two pure strategies, Hawk and Dove, are possible. Let the ESS be $P^*$ Hawk, $1-P^*$ Dove. An infinite random-mating diploid population plays this game. The choice of strategy is determined by two alleles at a locus, as follows:

| Genotype | 11 | 12 | 22 |
|---|---|---|---|
| Probability of Hawk phenotype | $P_0$ | $P_1$ | $P_2$. |

If the frequency of allele 1 is $p$, and of allele 2 is $q$, the frequency of the Hawk strategy is

$$F = p^2P_0 + 2pq\, P_1 + q^2P_2. \tag{4.1}$$

Writing $E(H)$ and $E(D)$ as the expected payoffs to Hawk and Dove against an opponent playing Hawk with frequency $F$, the genotypic fitnesses are

$$\begin{aligned} W_{11} &= P_0E(H) + (1-P_0)\, E(D) \\ W_{12} &= P_1E(H) + (1-P_1)\, E(D) \\ W_{22} &= P_2E(H) + (1-P_2)\, E(D). \end{aligned} \tag{4.2}$$

The search for the genetic equilibria is helped by a result of Lloyd (1977), who showed that if only two phenotypes are possible, and if genetic variability is caused by two alleles at a locus, then a genetic polymorphism exists only if:

*either* (i) the fitnesses of the two phenotypes are equal; that is, at the ESS frequency, when $F = P^*$,

*or* (ii) the relative frequencies of the two alleles are the same in the two phenotypes.

In case (ii), with unequal fitnesses, it is easy to show that the frequency of allele 1 at equilibrium is

$$\hat{p} = (P_1 - P_2)/(2P_1 - P_0 - P_2) \tag{4.3}$$

The existence of such an equilibrium, with $0 < \hat{p} < 1$, requires that $P_1 - P_0$ and $P_1 - P_2$ have the same sign; i.e. there is overdominance. It follows that, if there is no overdominance, the population either becomes genetically homozygous or it reaches the ESS. Which of these occurs depends on whether $P^*$ lies between $P_0$ and $P_2$. The various possibilities are illustrated in Figure 5*a*. If $P^*$ lies within the genetically possible range, the population will be genetically polymorphic with the Hawk and Dove phenotypes at the ESS frequencies.

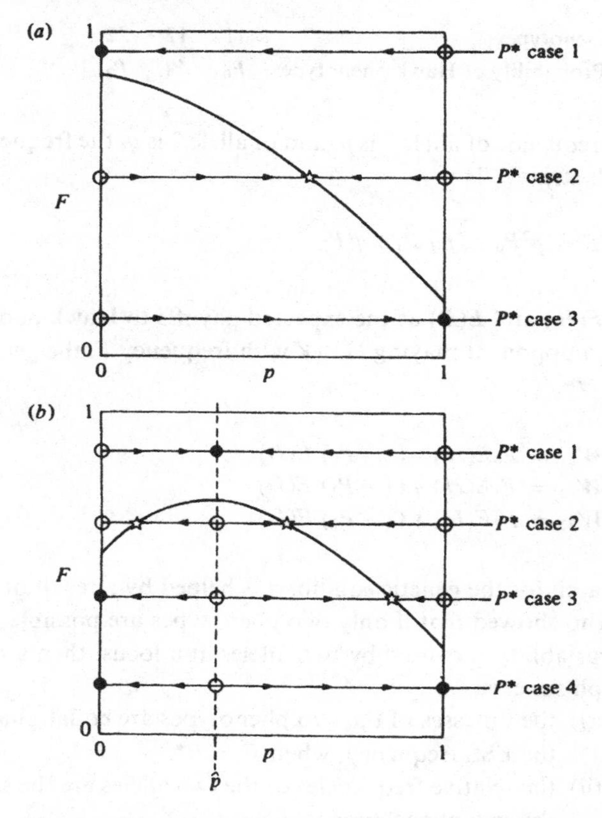


Figure 5. The Hawk–Dove game with diploid inheritance; (*a*) no overdominance; (*b*) with overdominance. $p$, gene frequency; $\hat{p}$, gene frequency such that $p$ is the same in Hawk and Dove phenotypes; $F$, frequency of Hawk in population; $P^*$, frequency of Hawk at ESS; ○, unstable equilibria; ●, stable equilibria; ☆, ESS's. The bold line is graph of $F$ against $p$.

If not, the population becomes fixed for the genotype closest to the ESS.

If there is overdominance things are more complex. The four possible cases are illustrated in Figure 5*b*. If $P^*$ lies within the genetically possible range it is always stable. If not, the population will usually evolve to a state as close to the ESS as possible.

For the two-pure-strategy game, then, the introduction of diploid genetics makes almost no difference. In more complex games, difficulties of two kinds can arise. First, as pointed out above, if the genetic system is simple there may be no way in which the ESS frequencies can be produced. Secondly, there is no guarantee that a genetic polymorphism $\hat{\mathbf{p}}$, corresponding to a mixed ESS $\hat{\mathbf{P}}$, will be stable. This difficulty is not peculiar to a sexual population; as explained in Appendix D, it applies also to asexual ones.

## B Phenotypes concerned with sexual reproduction

If we are concerned with such aspects of the phenotype as the sex ratio, sexual investment or anisogamy, we cannot ignore sexual reproduction. We can, however, use the basic idea of an uninvadable strategy to investigate such problems. It is natural to refer to such strategies as ESS's, although usually they cannot be found by applying conditions (2.4*a*,*b*). In this section I show how such problems can be tackled by analysing an example from sex ratio theory.

Fisher (1930) argued that, if the sex ratio was under parental control, the only stable state would be a 1 : 1 ratio, or, more generally, equal expenditure on sons and daughters. The reason is that if one sex is more common than the other it will pay parents to produce only the rarer sex. Although not couched in game-theoretic terms, his argument is, essentially, to seek an ESS, as was Hamilton's (1967) search for an 'unbeatable strategy' for the sex ratio with local competition for mates.

Let us, then, tackle Fisher's problem of the evolutionarily stable sex ratio in a random-mating population, but for a more general assumption about the 'phenotype set', or set of possible strategies. Thus suppose that the set of possible families lies within a phenotype set, as shown in Figure 6. The sex ratio is determined by genes in the mother; an exactly similar conclusion would emerge if it were controlled by genes in the father. Let the evolutionarily stable sex ratio be $a^*$ males : $b^*$ females in a family. Clearly, $a^*b^*$ will lie on the boundary of the phenotype set.

Consider a dominant mutant, $m$, which alters the sex ratio as shown in Table 8. The problem is to find $a^*b^*$ such that no mutant can invade. We need consider only mutants lying on the boundary of the set.

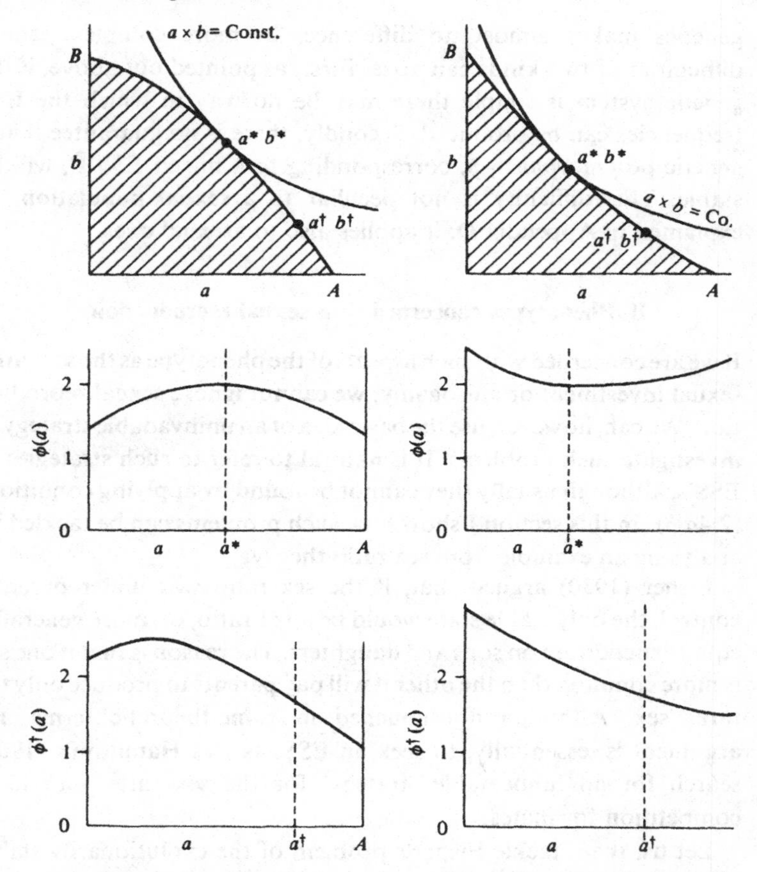


Figure 6. The sex ratio game. The three diagrams on the left refer to a convex phenotype set, and on the right to a concave phenotype set. The top two diagrams show the fitness sets; the shaded area represents the set of possible families. The central diagrams show $\phi(a) = a/a^* + b/b^*$ as a function of $a$. The bottom diagrams show $\phi\dagger(a) = a/a\dagger + b/b\dagger$ as a function of $a$. $a$, number of sons; $b$, number of daughters; $A$, $B$, maximum possible numbers of sons and daughters in single-sex families; $a^*$, $b^*$, values maximising the product $a \times b$; $a\dagger$, $b\dagger$, an alternative point on the boundary of the phenotype set.

Let the frequency of $+/m$ males be $p$ and of $+/m$ females be $P$. Since $m$ is small and mating is random, we can ignore $m/m$ genotypes and $+/m \times +/m$ matings. We can now write down the frequencies of

Table 8. *Genetic model of the evolution of the sex ratio*

| Genotype of parent | | Number of offspring | |
|---|---|---|---|
| Mother | Father | Sons | Daughters |
| $+/+$ | Any | $a^*$ | $b^*$ |
| $+/m$ or $m/m$ | Any | $a$ | $b$ |

Table 9. *The evolution of the sex ratio*

| Type of mating | | Frequency | Number of offspring | | | |
|---|---|---|---|---|---|---|
| | | | Male | | Female | |
| Mother | Father | | $+/m$ | $+/+$ | $+/m$ | $+/+$ |
| $+/m$ | $+/+$ | $P$ | $a/2$ | $a/2$ | $b/2$ | $b/2$ |
| $+/+$ | $+/m$ | $p$ | $a^*/2$ | $a^*/2$ | $b^*/2$ | $b^*/2$ |
| $+/+$ | $+/+$ | $1-P-p$ | — | $a^*$ | — | $b^*$ |

the three kinds of matings, and of the offspring produced, as in Table 9. Writing $p'$ and $P'$ as the frequencies of $+/m$ males and females, respectively, in the next generation, we have:

$$p' = \tfrac{1}{2}\frac{a}{a^*}P + \tfrac{1}{2}p \qquad (4.4a)$$

$$P' = \tfrac{1}{2}\frac{b}{b^*}P + \tfrac{1}{2}p. \qquad (4.4b)$$

Adding these equations gives

$$(p' + P') = p + P + RP$$

where
$$R = \tfrac{1}{2}\left(\frac{a}{a^*} + \frac{b}{b^*}\right) - 1. \qquad (4.5)$$

Note that if $a = a^*$, $b = b^*$ then $R = 0$. That is, if the mutant does not alter the phenotype, there is no change in $p+P$; this merely confirms that no mistake has been made in writing down equations (4.4*a*,*b*).

We seek values $a^*$ and $b^*$ such that $R<0$ for any $ab$ mutant not identical to $a^*b^*$. If we can find such an $a^*b^*$, it will be uninvadable.

Let $a/a^*+b/b^* = \phi(a)$, and let $f(a)$ be the boundary of the phenotype set. Then, if we consider a series of points lying on the boundary of the set, stability requires that $\phi$ be a maximum when $a = a^*$, $b = b^*$. Thus if $\phi(a) > \phi(a^*)$ for any $a$, an $ab$ mutant could invade an $a^*b^*$ population. Hence, for stability

$$[\partial\phi(a)/\partial a]_{a=a^*} = 0, \tag{4.6a}$$

and

$$[\partial^2\phi(a)/\partial a^2]_{a=a^*} < 0. \tag{4.6b}$$

Condition (4.6*a*) gives $b^*+a^*f'(a) = 0$, which is satisfied at that point on the boundary at which the product $a \times b$ is maximised, as shown in Figure 6. This is a more formal derivation of the result obtained by MacArthur (1965).

Note that equation (4.6*a*) only guarantees stationarity, not stability. For stability, condition (4.6*b*) gives

$$[\mathrm{d}^2\phi(a)/\mathrm{d}a^2]_{a=a^*} < 0.$$

That is, the phenotype set must be convex. Strictly, this condition only requires that the set be locally convex near $a^*b^*$, and only guarantees stability against mutants of small phenotypic effect. Global stability is best analysed by graphical methods.

The upper diagrams in Figure 6 show convex (*left*) and concave (*right*) phenotype sets, showing the stationary point $a^*b^*$ which maximises $a \times b$ on each of them, and also an alternative point, $a\dagger b\dagger$, on the boundary. In the centre row of diagrams in Figure 6, $\phi(a)$ is plotted against $a$. For the convex set, $\phi(a)$ is a maximum when $a = a^*$, and no value of $a$, with $0<a<A$, gives a value of $\phi(a)$ greater than $\phi(a^*)$. This means that a population at $a^*b^*$ is uninvadable by any mutant. For the concave set, $\phi(a)$ is a minimum when $a = a^*$. A population $a^*b^*$ could be invaded by *any* mutant on the boundary. In particular, it could be invaded by females producing offspring of only

one sex, all males or all females. For the concave set, it is easy to show that the stable state of the population consists of equal numbers of male-producing ($A$,0) and female-producing (0,$B$) females.

To complete these two cases of convex and concave sets, the lower diagrams in Figure 6 show $\phi\dagger(a) = a/a\dagger + b/b\dagger$, as a function of $a$. Note that, whatever the form of the fitness set, there are always values of $a$ for which $\phi\dagger(a) > \phi\dagger(a\dagger)$. This means that there are always mutants which can invade an $a\dagger b\dagger$ population.

It is natural to refer to the phenotype $a^*b^*$ in Figure 6 (*top*) as an ESS; it is drawn from a set of possibilities defined by the phenotype set, and it has the property of being uninvadable by any mutant. However, the ESS can no longer be derived from conditions (2.4*a*,*b*); instead, one must write down recurrence relations in terms of gene frequencies or, if mating is not random, in terms of the frequencies of the different kinds of matings.

In the particular problem just treated, we could find from the recurrence relations a function $R$ which was maximum at the ESS. It is not always obvious how to do this. There is, however, a general method for finding the ESS from recurrence relations of this kind; this method is described in Appendix I.

### C The evolution of anisogamy

The origin of anisogamy is a problem which, self-evidently, requires that the diploid sexual method of reproduction be taken into account, but which, nevertheless, can most easily be analysed as a game (Maynard Smith, 1978). The basic hypothesis was formed by Parker, Baker & Smith (1972), who pointed out that the origin of anisogamy can be explained selectively if the following assumptions are made:

(i) Reproduction is essentially sexual, involving the production of gametes by meiosis, followed by the fusion of gametes to form a zygote.

(ii) The probability that a zygote will survive to become a gamete-producing adult is an increasing function of the mass of the zygote.

(iii) The total mass of gametes produced is limited to some value $M$, but $M$ can be divided to produce many small gametes or a few large ones.

Of course, 'mass' in assumptions (ii) and (iii) could be replaced by any limiting resource: for example, protein content.

The first assumption makes it explicit that we are not attempting to answer the harder question of why organisms reproduce sexually in the first place; instead, sexual reproduction is assumed, and the selective forces favouring anisogamy as opposed to isogamy are analysed. It is convenient to make a fourth assumption (Maynard Smith, 1978):

(iv) There is a minimum mass, $\delta$, below which a cell cannot function effectively as a gamete.

If this fourth assumption is not made, one can be led to the unrealistic conclusion that a parent should produce an infinite number of gametes of zero mass. Given these assumptions, it turns out that the evolution of isogamy or anisogamy depends only on the form of the graph relating survival probability to zygote size.

Before developing the model, two points need discussing. First, an alternative scenario for the evolution of anisogamy has recently been proposed by Cosmides & Tooby (1981). They suggest that gamete size might be controlled by cytoplasmic as well as nuclear genes. If so, it becomes relevant that, at least in some cases, if two gametes carrying different cytoplasmic genes fuse, then one of the alleles may increase in number of copies relative to the other in the resulting zygote. The successful allele is the one which was initially most abundant in the newly-formed zygote (Birky, 1978). Consequently, cytoplasmic genes which caused the production of large gametes would be favoured, and this could lead to the production of gametes of different sizes. The idea is an interesting one which deserves further investigation, but it will not be pursued here.

The second point to be discussed is a criticism of Parker *et al.*'s proposal which has been made by Wiese, Wiese & Edwards (1979). These authors point out that, in all probability, gamete bipolarity is older than anisogamy. In most sexual isogamous organisms, gametes, although morphologically indistinguishable, are of two kinds, + and −, such that fusion takes place only between unlike types. I accept this, but do not think it alters the problem substantially. It does, however, remove a difficulty discussed by Parker (1978). Why, once anisogamy has arisen, do not macrogametes fuse with each other rather than with microgametes? The answer may be that they do not fuse because they are of the same mating type. Parker could, of course, reply that one must still explain why mating type incompatibility does not break down in anisogamous organisms.

In what follows, I accept Wiese *et al.*'s argument, and assume the existence of two mating types, both in isogamous and anisogamous populations. Indeed, their data are worth reviewing before we start the theoretical analysis. In three species of *Chlamydomonas*, *C. reinhardti*, *C. chlamydogama* and *C. moewusii*, gametes are typically produced by two (or three) mitotic divisions of a haploid cell, giving rise to four (or eight) small biflagellate gametes. In all three species, however, larger gametes can also be produced without division from non-dividing cells. All gametes produced by a given clone are of the same mating type, but can be of different sizes. Anisogamous fusion can therefore occur, between large + gametes and small – gametes, or vice versa. Other *Chlamydomonas* species are typically anisogamous, with different clones producing large and small gametes. In the extreme form, *C. pseudogigantea*, the large gametes are unflagellated.

These observations suggest the following evolutionary stages:

(i) All clones produce morphologically identical gametes; some clones produce only + and some only – gametes.

(ii) All clones produce both large and small gametes, but all gametes produced by a clone are of the same mating type.

(iii) Some clones specialise in producing large and some in producing small gametes.

(iv) Further differentiation in gametes, and in secondary sexual characters.

This is essentially a dioecious scenario; Wiese (1981) outlines an alternative, monoecious, road to anisogamy.

Turning now to a mathematical model with assumptions (i) to (iv) above, let $S(x)$ be the probability that a zygote of size $x$ will survive to become a breeding adult. $S(x)$ will be an increasing function of $x$, with a maximum value of 1 and a minimum value $S(0) = 0$. An adult can produce $n$ gametes each of size $m$, subject to the constraint

$$nm = M, \tag{4.7}$$

where $M$ is a constant.

With these assumptions, we can now ask some questions:

> *Would a population producing the smallest possible gametes be evolutionarily stable?*

In a population of individuals producing gametes of size $\delta$, a typical individual produces $M/\delta$ gametes. The typical zygote mass is $2\delta$. Hence, if we measure fitness by the number of offspring surviving to become adults, the fitness of a typical individual is

$$W_\delta = \frac{M}{\delta} S(2\delta).$$

Consider now a mutant producing gametes of size $m$, where $m > \delta$. These gametes would fuse with gametes of size $\delta$, of opposite mating type. Hence

$$W_m = \frac{M}{m} S(m+\delta).$$

So a $\delta$-producing population is evolutionarily stable providing that, for all $m > \delta$,

$$\frac{M}{\delta} S(2\delta) > \frac{M}{m} S(m+\delta)$$

or

$$\frac{S(2\delta)}{\delta} > \frac{S(m+\delta)}{m}. \tag{4.8}$$

Figure 7*a* shows a form of the function $S$ for which expression (4.8) holds, and 7*b* one for which it does not. I believe that Figure 7*a* represents the primitive condition, of an isogamous population in which gametes are as small as is compatible with their functioning effectively as gametes. Males were the first sex.

As adult size increased relative to $\delta$, a survival function of the type in Figure 7*a* might be replaced by that in Figure 7*b*. At this point, a clone of macrogamete-producers could invade. Would this lead to anisogamy, or to a new type of isogamy, this time of macrogamete-producers? To answer this question, we first seek a size, $m^*$, of a macrogamete which would be stable against invasion by mutants of

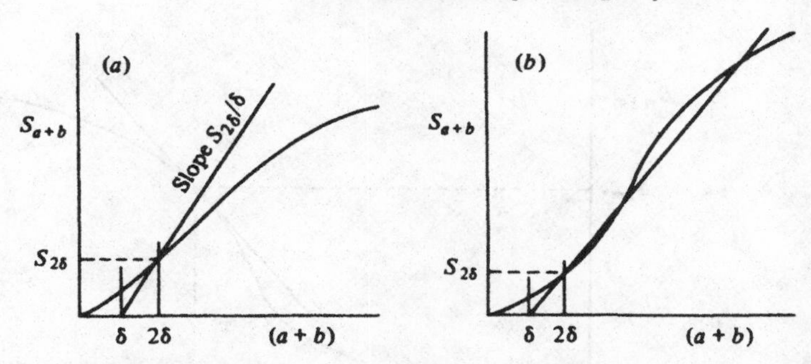


Figure 7. Zygote survival, $S$, as a function of zygote size, $a+b$. ($a$) Production of microgametes is evolutionarily stable; ($b$) production of microgametes is not stable. $\delta$, minimum possible gamete size.

small phenotypic effect, and then ask whether a population of $m^*$-producers could be invaded by $\delta$-producers. Thus we ask:

*What is $m^*$, the locally stable gamete size?*

Consider a mutant clone producing gamete of size $m$ in a population of $m^*$-producers. The fitness of an $m$-producer is then

$$W_m = \frac{M}{m} S(m+m^*).$$

To find the ESS, we adopt the method outlined in Appendix I. Thus,

$$\frac{\partial W_m}{\partial m} = \frac{M}{m} \frac{\partial S(m+m^*)}{\partial m} - \frac{M}{m^2} S(m+m^*),$$

and if $m^*$ is an ESS, $W_m$ must be a maximum when $m = m^*$; that is

$$(\partial W_m/\partial m)_{m=m^*} = 0,$$

or
$$[\partial S(m+m^*)/\partial m]_{m=m^*} = S(2m^*)/m^*. \tag{4.9}$$

This condition is shown graphically in Figure 8. Note that there need not be a locally stable value of $m^*$. Such a value will exist only if, at some point, the slope of $S(x)$ against $x$ is greater than $2S(x)/x$.

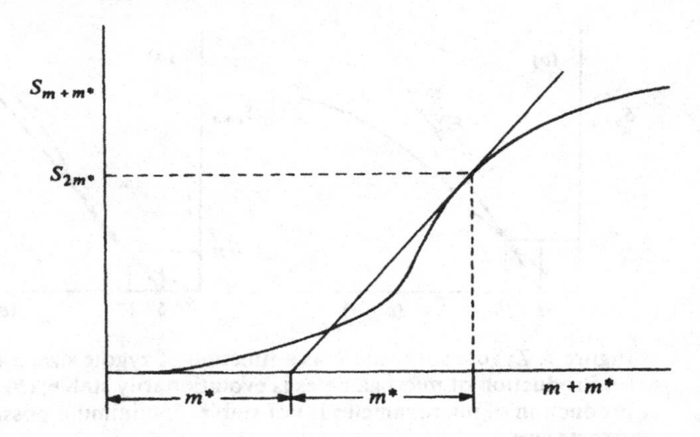


Figure 8. Conditions under which the production of gametes of size $m^*$ is evolutionarily stable against mutations of small phenotype effect. For other symbols, see text.

Suppose, however, that such a locally stable value, $m^*$, does exist. The question remains:

> *Would a population of $m^*$-producers be stable against invasion by a $\delta$-producer?*

Stability of $m^*$ requires that $W_{m^*} > W_\delta$; that is,

$$\left.\begin{aligned} &\frac{M}{m^*}S(2m^*) > \frac{M}{\delta}S(m^*+\delta), \\ \text{or} \quad &\frac{S(2m^*)}{m^*} > \frac{S(m^*+\delta)}{\delta}. \end{aligned}\right\} \quad (4.10)$$

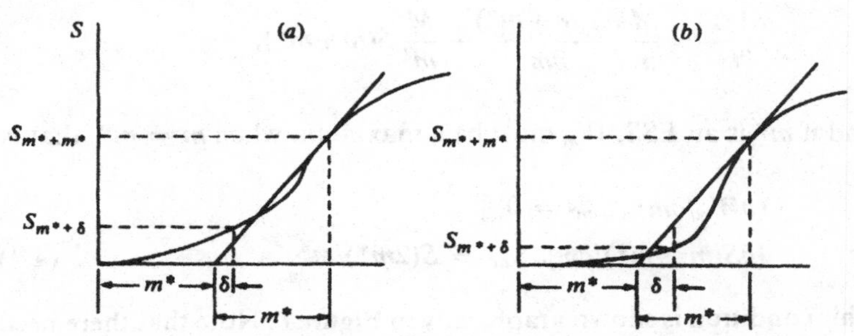


Figure 9. Survival curves for which macrogamete production is (*a*) unstable and (*b*) stable. Symbols as for Figures 7 and 8.

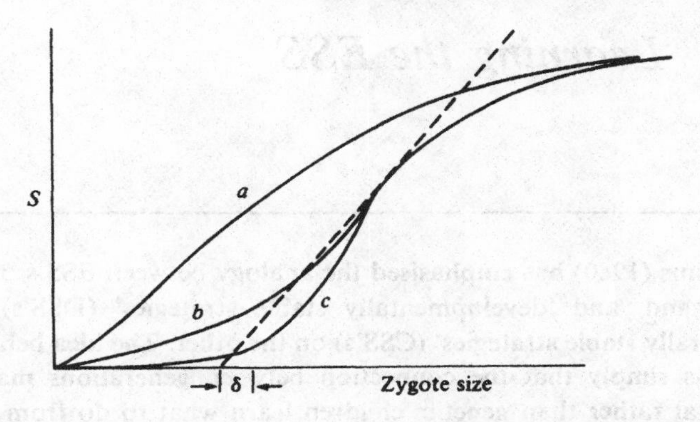


Figure 10. Three curves of survival probability, *S*, against zygote size. (*a*) Microgamete production stable; (*b*) anisogamy stable; (*c*) macrogamete production stable. $\delta$, minimum possible gamete size.

This condition is illustrated in Figure 9*b*. The form of $S(x)$ is not a very plausible one; for $m^*$ to be stable against $\delta$, it must be the case that, if $2m^*$ is the optimum size of a zygote, then zygotes of half that size have an almost zero probability of survival. Given the more plausible form of $S(x)$ in Figure 9*a*, $\delta$-producers can invade. If, in addition, condition (4.8) is untrue, so that a population of $\delta$-producers can be invaded, then microgamete- and macrogamete-producers will coexist; the population will be anisogamous.

Figure 10 shows three forms of $S(x)$; 10*a* gives isogamy for microgamete production, and is, I believe, primitive; 10*b* gives anisogamy; 10*c*, which would select for isogamy for macrogamete production, may never arise in practice.

# 5 *Learning the ESS*

Dawkins (1980) has emphasised the analogy between ESS's on the one hand, and 'developmentally stable strategies' (DSS's) and 'culturally stable strategies' (CSS's) on the other. The idea behind a CSS is simply that the connection between generations may be cultural rather than genetic: children learn what to do from their parents, or other members of the previous generation, instead of inheriting genes which determine their behaviour. Cultural inheritance has been analysed formally by Feldman & Cavalli-Sforza (1976) and by Lumsden & Wilson (1981); it is considered from a game-theoretic point of view in Chapter 13. This chapter is concerned with developmentally stable strategies, in which the analogue of genetic inheritance is learning, not cultural transmission.

The example which Dawkins uses to illustrate the idea of a DSS is so apt that I cannot resist borrowing it. Baldwin & Meese (1979) studied the behaviour of a pair of pigs in a Skinner box, arranged so that when a lever at one end of the box was pressed, food was dispensed at the other. They found that in those cases in which the pair developed a stable pattern of behaviour, the dominant pig pressed the bar and then rushed over to the food dispenser, while the subordinate pig waited at the dispenser. Such behaviour is stable for the following reason. Provided that enough food is dispensed at each press of the bar to ensure that some is left when the dominant pig arrives, the dominant is rewarded for bar-pressing; obviously, the subordinate is rewarded for waiting at the dispenser. The reverse pair of behaviours would not be stable; the subordinate would not be rewarded for pressing the bar because the dominant would prevent it from eating. Paradoxically, the observed behaviour is stable even if the quantity of food dispensed is such that the subordinate gets more of it than the dominant.

It would be easy to construct a hypothetical example in which analogous behaviour was genetically determined instead of learnt.

This analogy between learning and evolution has been investigated further by Harley (1981), on whose work the rest of this chapter is based. The problem is complex, because there is not only a formal analogy between learning and evolution; there is also a causal connection between them. Learning evolves, and we can therefore ask what kinds of learning rules will be evolutionarily stable. Harley's central result is that, for a rather general model, the learning rules which will evolve are precisely those which will, within a generation, take a population to the ESS frequencies.

Suppose that during its lifetime an animal plays a number of games, each one many times, just as a man might play chess, tennis and solitaire, each many times. Thus an animal plays the 'foraging game', the 'mating game', the 'peck-order game', and so on. We consider only games which are played often, because there can be no learning of a game played only once. The games can be of different kinds, as follows:

(i) Frequency-independent; i.e. 'games against nature', in which the payoff to a strategy is independent of the frequency with which it is played.

(ii) Frequency-dependent.

- (*a*) Individual games; i.e. games in which the payoff does not depend on what other members of the population are doing, but does depend on the frequencies with which the individual adopts different actions.
- (*b*) Population games; i.e. games in which the payoffs do depend on what other members of the population are doing.

Although the logic of these different kinds of games varies, there is no reason to suppose that an animal knows which kind of game it is playing. Thus an animal knows whether it is foraging, mating or competing for dominance, but not whether the activity in question is frequency-dependent in its payoffs. In less anthropomorphic terms, a foraging animal will modify its behaviour in the light of previous experience when foraging, but not when mating, but it will use the same rules for modifying – i.e. the same 'learning rule' – in both cases.

Population games can take various forms. In Harley's model it is assumed that individuals pair off at random, engage in a single contest, adjust their strategies in the light of the payoffs received and their learning rule, and then pair randomly with a new opponent, continuing the process until the distribution of strategies reaches a steady state. As in the evolutionary case, the dynamics are similar if an individual is 'playing the field'; an example is the game of 'digging' versus 'entering' in digger wasps, studied by Brockmann *et al.* (1979; see p. 74), although, surprisingly, it seems that in that game individuals are not learning.

I suspect that Harley's model will apply more to games against the field than to pairwise contests, for the following reason. In a pairwise contest, such as that between Baldwin and Meese's pigs, individuals will usually play against the same opponent several times. If individual recognition is possible, an animal can treat contests against different opponents as different games, and develop a different strategy for each one. For the present, therefore, population games are supposed to be either against the field, or against a random series of opponents. Later (p. 66), when a learning rule appropriate for such population games has been found, I will discuss what happens if two animals, each adopting that learning rule, engage in a series of contests with each other.

First, we must distinguish between two concepts, an 'ES learning rule' and a 'rule for ESS's'.

An 'evolutionarily stable learning rule', or 'ES learning rule' for short, is a rule such that a population of individuals adopting that rule cannot be invaded, in evolutionary time, by mutants adopting different learning rules. It is, therefore, a learning rule which is evolutionarily stable in exactly the same sense that any other strategy might be an ESS; it satisfies equations (2.9) for some defined set of alternative learning rules.

A 'rule for ESS's' is a rule which, for some particular game or set of games, will take an initially naïve population to the ESS frequencies in a single generation of learning.

The first point to establish is that a rule for ESS's will also be a rule which takes an individual to an optimal strategy when playing a frequency-independent game with the same set of pure strategies. Thus imagine a population of individuals, all with the same learning rule, playing a game with the pure strategy set $A, B, C \ldots$. Suppose that the population is not at an ESS, but is in a state such that strategy

$A$ pays better than any other. If the learning rule is to bring the population to the ESS, it must be such as to increase the chance of adopting $A$ in the next round. Clearly, such a learning rule will take a population to the optimal strategy for a frequency-independent game.

We are now ready to tackle Harley's theorem, which asserts that an ES learning rule, if one exists, is necessarily also a rule for ESS's. To prove this, we must make a number of assumptions:

(i) During its lifetime, an animal plays one or a number of different games, which may be frequency-independent games against opponents, or frequency-dependent or frequency-independent games against nature.

(ii) All games played have an ESS (or an optimal solution).

(iii) Each game is played a large number of times, so that the payoff received after a steady state is reached overrides in importance the payoff received while learning.

(iv) At least one rule for ESS's exists for the set of games.

Let $I$ be a learning rule which does *not* take the population to the ESS of all the games; it does not do so for games 1,2,3,. . . and it does do so for games $n$, $(n+1)$. . . . Consider a population of animals adopting rule $I$ invaded by a mutant $J$, which *is* a rule for ESS's (note that, by assumption (iv), at least one such rule exists). In games 1,2,3 . . . the population will reach some state which is not the ESS; it follows that there is some action $X$ which has a higher payoff than is being achieved by typical members of the population. To the mutant $J$, this presents itself as a frequency-independent game with optimal strategy $X$; as we have seen, the mutant $J$ will learn to play $X$. Hence, in games 1,2,3 . . ., once a steady state is reached, $J$ will do better than $I$, and in games $n$, $(n+1)$ . . ., $I$ and $J$ are equally fit. Hence $J$ is fitter than $I$, so that $I$ cannot be an ESS.

It follows that, if an ES learning rule exists, it must be a rule for ESS's.

We now have to qualify this statement. The ES learning rule will not lead to the complete loss or fixation of particular behaviours. Thus suppose that, for a particular game, the ESS is 'always do $A$'. Then behaviour $A$ should become genetically fixed; there is no point in learning. If learning has been retained, it is presumably because payoffs change in time or space. Thus the prediction that the ES learning rule takes the population to the ESS should be modified, by adding 'except when the ESS is a pure strategy, or does not include certain actions, in which case the rule will cause individuals to perform these excluded actions with low frequency'. As Darwin advised, one should occasionally do a fool's experiment, just in case.

Having established certain properties an ES learning rule must possess, we can now say something about what kind of rule it must be.

For each game, an animal has a set of possible behaviours, or actions, $B_i (i=1,2 \ldots n; n \geqslant 2)$. $P_i(t)$ is the payoff (change of fitness) an animal receives on trial $t$ for action $B_i$; if some action other than $B_i$ is made on trial $t$, then $P_i(t)=0$. The learning rule defines for each game the $n$ probabilities, $f_i(t)$, of action $B_i$ on trial $t$, as a function of the previous payoffs $P_i(\tau)$, where $\tau < t$.

Harley proves the following proposition. When a population with an ES learning rule has reached an equilibrium, the probability of performing action $B_i$ is equal to the total payoff to date received when performing action $B_i$ divided by the total payoff received for all actions. More formally

$$f_i(t)_{t \to \infty} \to \sum_{\tau=1}^{t-1} P_i(\tau) \Big/ \sum_{i=1}^{n} \sum_{\tau=1}^{t-1} P_i(\tau). \tag{5.1}$$

Note that the probability of action $B_i$ is given by the ratio of the *total* payoffs, not by the ratio of the *rates* of payoff per action.

The proof is as follows. Let $t_i$ = total number of times $B_i$ is adopted in $t$ trials ($t = \Sigma t_i$), and let $E[P_i(t)]$ = expected payoff to $B_i$ on trial $t$ given that $B_i$ is adopted. In a sufficiently long series of trials,

$$f_i(t) \to t_i/t,$$

and

$$E[P_i(t)] \to \sum_{\tau=1}^{t-1} P_i(\tau)/t_i.$$

From the Bishop–Cannings theorem (see Appendix C), $E[P_i(t)] = E[P_j(t)]$ for all $i,j$ in the support of the ESS. If this constant expected payoff is $C$, then

$$\sum_{\tau=1}^{t-1} P_i(\tau) \to t_i C$$

and hence

$$f_i(t) \to \sum_{\tau=1}^{t-1} P_i(\tau)/Ct.$$

Since

$$\sum_{i=1}^{n} f_i(t) = 1,$$

we have

$$Ct = \sum_{i=1}^{n} \sum_{\tau=1}^{t-1} P_i(\tau),$$

and expression (5.1) is proved.

To make this expression more meaningful, imagine a population playing a game with three pure strategies, of which the ESS is $\frac{2}{3}A$, $\frac{1}{3}B$, $0C$. At equilibrium, animals will adopt $A$ twice as often as $B$, and $C$ not at all. Since they are at the ESS, their expected payoffs for $A$ and $B$ are equal. Hence the total payoff from $A$ will be twice that from $B$. Applying expression (5.1) then tells us that the learning rule will leave the probabilities unaltered, as it should since the population is at the ESS.

Condition (5.1) describes behaviour when the ESS has been reached. It cannot itself be the ES learning rule, if only because it does not specify how a naïve animal should act. Harley suggests the following as an approach to a realistic ES learning rule:

$$f_i(1) = r_i \Big/ \sum_{i=1}^{n} r_i,$$

$$f_i(t) = \frac{r_i + \sum_{\tau=1}^{t-1} m^{t-\tau-1} P_i(\tau)}{\sum_{i=1}^{n} \left[ r_i + \sum_{\tau=1}^{t-1} m^{t-\tau-1} P_i(\tau) \right]}, \qquad (5.2)$$

where $0 < m < 1$.

In these equations the $r_i$ are the 'residual values' associated with each behaviour; if, for example, all the $r_i$ were equal, then all behaviours would be equally likely in the first trial. The closer $m$, a

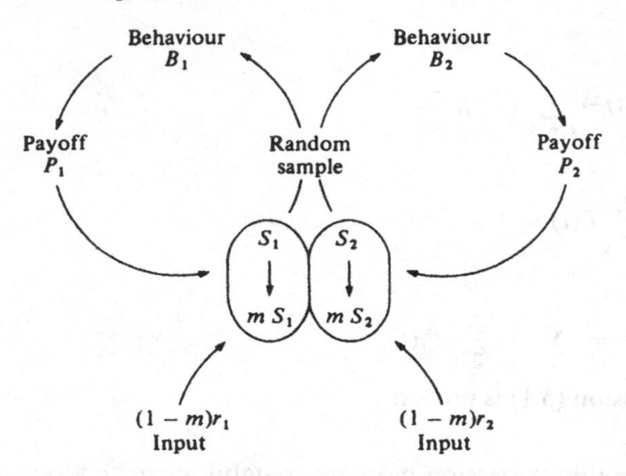


Figure 11. Mechanism of the 'relative payoff sum' (RPS) learning rule. For explanation, see text. (After Harley, 1981.)

memory factor, is to unity, the more attention an animal pays to earlier payoffs.

The nature of this 'relative payoff sum' (RPS) learning rule can be further clarified in two ways. First, a simple mechanism which could generate such a rule, and which could easily be realised in chemical or neuronal hardware, will be described. The point of doing this is partly to show that it is not unreasonable to suppose that a rule of this kind could exist, and partly to make it easier for those with a mechanical rather than a mathematical turn of mind to see what is happening. Secondly, the behaviour of animals adopting the rule will be simulated in various types of game.

Figure 11 shows a possible mechanism for the RPS learning rule, for a game with two possible behaviours, $B_1$ and $B_2$. At each trial, the choice of action depends on the concentrations $S_1$ and $S_2$ of some substance in two cells; the probabilities of choosing $B_1$ and $B_2$ are $S_1/(S_1+S_2)$ and $S_2/(S_1+S_2)$, respectively. For example, the cells could be neurones whose firing rates are proportional to the concentrations of some substance, and the choice of action could depend on which neurone fired first after some arbitrary instant. As a result of action $B_1$, a quantity $P_1$ (equivalent to payoff) is added to cell 1. In each time interval between trials, the quantities $S_1$ and $S_2$ are reduced, according to first-order chemical kinetics, to $mS_1$ and $mS_2$, respectively, and are increased by synthesis by amounts $(1-m)r_1$ and $(1-m)r_2$, respectively. Thus $r_1$ and $r_2$ represent an unlearnt bias in favour of actions $B_1$ or $B_2$; in the absence of payoffs, the concentrations will be $r_1$ and $r_2$, respectively.

Whether this model should be regarded as an invitation to neurophysiologists to seek for a physical realisation of a learning rule is a matter of judgement. The chemical kinetics assumed are simple and plausible. For our present purpose, however, it is sufficient that the model should make clear the nature of the RPS learning rule defined by expression (5.1).

Before describing simulations of this model, we can list some properties expected of it:

(i) It will take a population to the ESS, subject to the constraint that, because of the residual inputs $r_i$, no behaviour can be completely lost. For a game with only two behaviours, the upper and lower limits of $f_i$ are approximately $m$ and $1-m$.

(ii) Recent trials have a greater effect on behaviour than earlier ones. The smaller $m$, the shorter is the memory of earlier trials.

(iii) Initial behaviour is entirely determined by the residual rates, $r_i$. A naïve animal can have a bias in favour of a particular behaviour. A more sophisticated rule would permit the residuals to be modified by experience.

(iv) The rate of change of behaviour depends on the relative magnitudes of the residuals, $r_i$, and the payoffs, $P_i$. If the residuals are larger, animals will be slow to modify their initial naïve behaviour; in contrast, if payoffs are large relative to residuals, animals will quickly settle down to adopting a particular behaviour with high probability. The biological significance is as follows. The residuals represent an initial expectation of payoffs for a particular game. If actual payoffs are smaller than this, an animal should continue to switch behaviours. If payoffs are higher than expectations, it should stick to the particular behaviour that has paid off, which is equivalent to changing its initial random behaviour.

Harley describes simulations of four games:

(i) *The two-armed bandit*. This is a frequency-independent game, in which there are two possible behaviours, $B_1$ and $B_2$, each with a constant probability of yielding a reward. The two probabili-

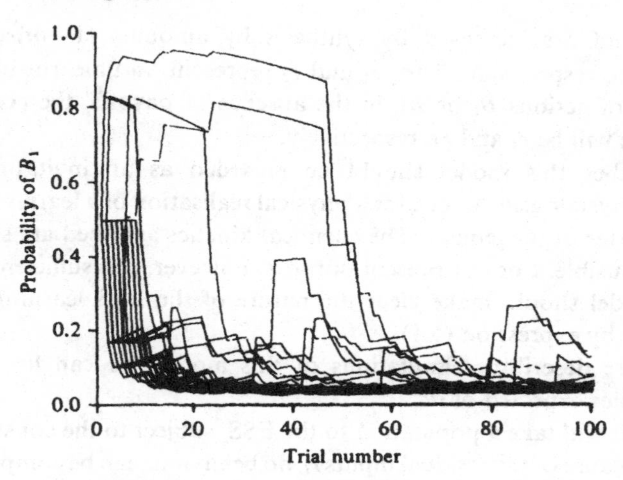


Figure 12. Simulation of the RPS learning rule playing the two-armed bandit. The probability of choosing $B_1$, the less profitable arm, is shown as a function of trial number for 30 replications of the game. The probabilities of rewards at the two arms were 0.1 and 0.4. Residuals $r_1$ and $r_2$ were 0.25 and $m = 0.95$. (After Harley, 1981.)

ties are initially unknown, but can be estimated by trial and error. The ESS, in a sufficiently long series of trials, is to adopt always the behaviour with the higher probability of payoff, once this is known. This is what is done in fact by the RPS learning rule, except that an animal never becomes completely fixed for the optimal behaviour. An example is shown in Figure 12. There is experimental evidence (Bush & Wilson, 1956, on paradise fish; Roberts, 1966, on rats; Krebs, Kacelnik & Taylor, 1978, on great tits) that vertebrates behave in this way.

(ii) *The Hawk–Dove game.* Figure 13 shows simulations of a population of 30 individuals playing the Hawk–Dove game. The population mean stays close to the ESS of 80% Dove. Individuals drift considerably, however, with some adopting predominantly Hawk and others Dove. This happens because expected payoffs at the ESS are equal, so there is no inducement to change, but some animals

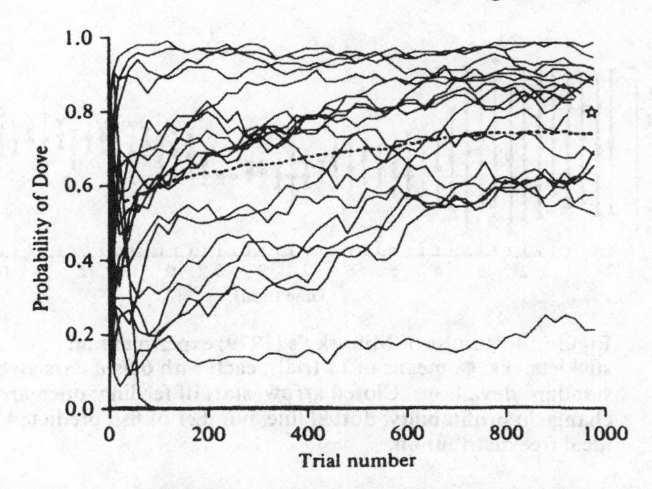


**Figure 13. Simulation of the RPS learning rule playing the Hawk–Dove game. Full lines, probability of playing Dove for 20 individuals, in a population of 30; broken line, population average. The ESS (☆) is $P(D) = 0.8$. (After Harley, 1981.)**

may, by chance, have initially fared better with one behaviour or the other.

(iii) *Population foraging*. This is a simulation of an experiment by Milinsky (1979) on sticklebacks. Six fish in a tank were fed with *Daphnia* from both ends of the tank, the rate being twice as great at one end as at the other. The ESS, when no fish could gain by moving to the other end of the tank, arises where there are four fish at the end with the higher rate of supply and two at the other. Figure 14 shows the results of such experiments. Statistically, the fish distribute themselves according to the ESS prediction, or, more precisely, according to the 'ideal free distribution' of Fretwell & Lucas (1970); this concept is discussed further on p. 90. Individual fish, however, continued to move from one end of the tank to the other, as is to be expected, since in nature the relative profitabilities of patches will not remain constant. In fact, Milinsky did switch the relative feeding rates between the two ends during the course of each experiment, and the fish redistributed themselves accordingly.

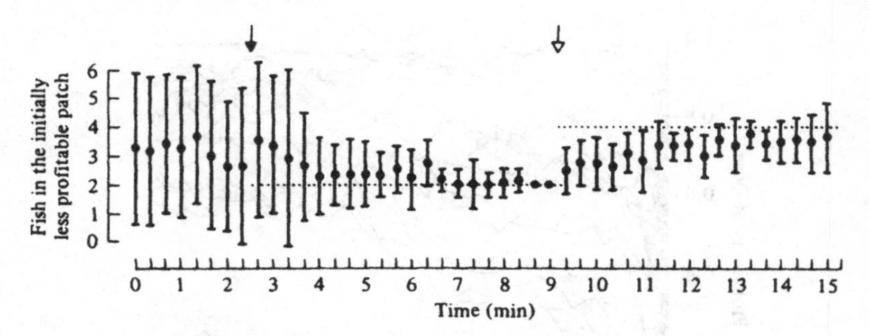


Figure 14. Results of Milinsky's (1979) experiment on sticklebacks. ●, means of 11 trials, each with 6 fish; bars give standard deviations. Closed arrow, start of feeding; open arrow, change in profitability; dotted line, number of fish predicted from ideal free distribution.

Harley simulated Milinsky's experiments, assuming the fish had an RPS learning rule, and obtained closely similar results. Figure 15 shows a further series of simulations, illustrating the effect of varying the values of the residuals relative to the expected payoffs. When residuals are small (Figure 15*b*), individual animals settle down rather quickly at one end or the other; when residuals are large (Figure 15*c*) individuals retain a high level of exploration. In all cases, however, the population mean gradually approaches the ESS. This figure illustrates a prediction which could be tested in an experimental setup such as Milinsky's. Hungry fish should resemble Figure 15*b*, and well-fed fish Figure 15*c*. Although Milinsky did not test this prediction, there are data in accord with it (Heller & Milinsky, 1979).

(iv) *The 'concurrent variable-interval' game*. This is a game commonly played between pigeons and experimental psychologists. As in the two-armed bandit, two choices are possible, say 'left' and 'right'. Each has a constant probability of delivering a food item on each trial. But if the animal tries the left arm in any trial, the right arm will also deliver a food item with its own probability, although the item is unavailable to the animal and cannot be seen. The item remains there, and can be eaten next time the animal tries the right arm. Once an item is available, no further item will be delivered on that side until it has been consumed.

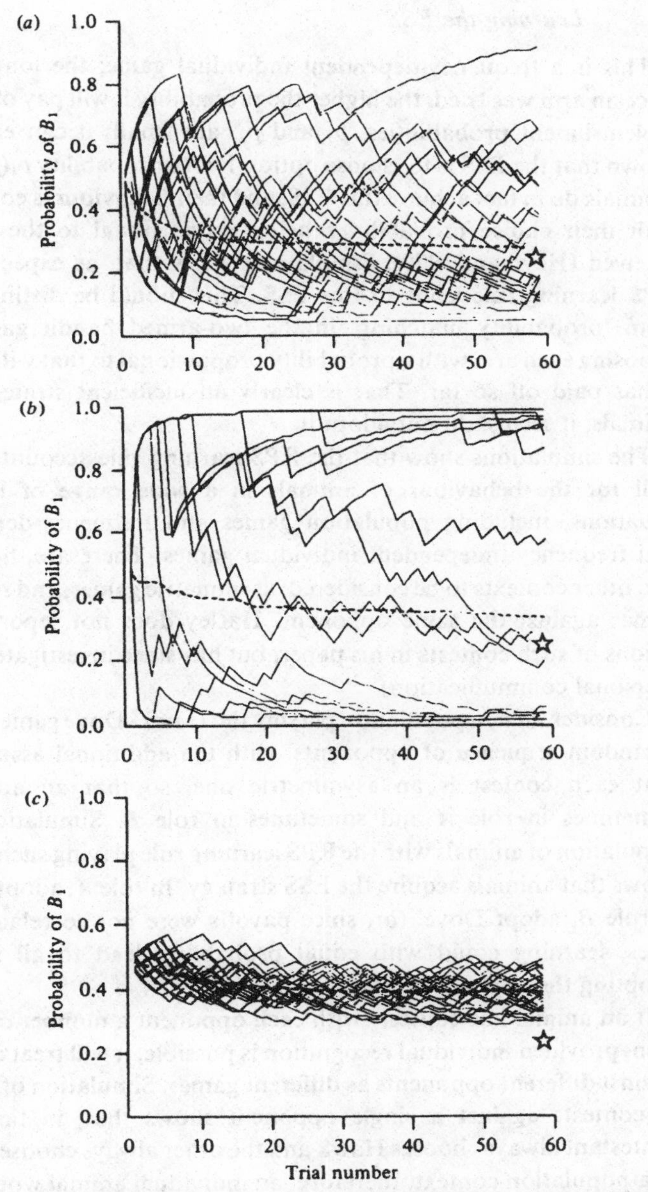


**Figure 15. Simulations of population foraging using the RPS learning rule. Full lines, individual means; broken line, population mean; ☆, ESS. The expected payoff rates, given by the residuals, were (*a*) intermediate, (*b*) small and (*c*) large, respectively. (After Harley, 1981.)**

This is a frequency-dependent individual game; the longer it is since an arm was tried, the higher the probability it will pay off. If the replenishment probabilities, $p_1$ and $p_2$, are small, it can easily be shown that the ESS is to choose option 1 with probability $p_1(p_1+p_2)$. Animals do in fact achieve this ESS, and their behaviour is consistent with their choice probabilities being proportional to the payoffs received (Heyman, 1979). Simulations show that, as expected, the RPS learning rule leads to this ESS. This should be distinguished from 'probability matching' in the two-armed bandit game; i.e. choosing each arm with a probability proportional to that with which it has paid off so far. That is clearly an inefficient strategy, and animals, it seems, do not adopt it.

The simulations show that the RPS learning rule accounts rather well for the behaviour of animals in a wide range of learning situations, including population games and frequency-dependent and frequency-independent individual games. There are, however, two other contexts to be considered: asymmetric games, and repeated games against the same opponent. Harley does not report simulations of such contests in his paper, but has since investigated them (personal communication).

Consider, first, a population playing the Hawk–Dove game against a random sequence of opponents, with the additional assumption that each contest is an asymmetric one, so that an animal is sometimes in role $A$ and sometimes in role $B$. Simulation of a population of animals with the RPS learning rule playing such a game shows that animals acquire the ESS strategy 'In role $A$, adopt Hawk; in role $B$, adopt Dove' (or, since payoffs were not correlated with roles, learning could with equal probability lead to all animals adopting the opposite strategy).

If an animal has contests with each opponent a number of times, then, provided individual recognition is possible, it will treat contests against different opponents as different games. Simulation of a series of contests against a single opponent shows that, in time, one contestant always chooses Hawk and the other always chooses Dove. In a population context, therefore, an individual animal would play Hawk against some opponents and Dove against others, but escalated contests would be rare because, for each pair, one would choose Hawk and one Dove.

One difficult problem remains: in proving that the ES learning rule is necessarily a rule for ESS's, it was assumed that payoffs correspond to changes in fitness. We can, if we like, make this true by definition; that is, we can define the payoff for a particular action as being equal to the change in expected number of offspring resulting from the action. We are then left with the problem of how the immediate consequences of some action can be translated by the animal into fitness units: equivalently, how can the synthesis of some chemical, in the model in Figure 11 (p. 60), be made proportional to a change in fitness? In ordinary evolutionary game theory, no such difficulty arises: it is precisely the change in fitness which causes the change in the relative frequencies of the phenotypes in the populations. In contrast, what a learning animal knows is whether it is hungry or thirsty or in pain, but not what effect this may have on its future reproduction.

Two general points can be made. First, the difficulty is not peculiar to learnt behaviour. If an animal's behaviour is to be appropriate to its survival, then, whether the behaviour is learnt or instinctive, there must be an appropriate translation of sensory input into motivational state, according to the utility of particular types of behaviour in fitness terms (McFarland, 1974). The second point is that there will be strong selection favouring animals which are most successful in performing such translations. An animal which performs the 'correct' action – correct in fitness-maximising terms – when simultaneously experiencing hunger, thirst and sexual motivation, will, by definition, leave most offspring. There is, of course, no conceivable translation system which can guarantee the correct action in all circumstances: moths fly into candles, human beings become addicted to heroin and reed warblers raise baby cuckoos. We can expect the best fit between theory and observation when all rewards are in the same currency, as would be the case, for example, in foraging theory if we could forget about the risks of predation.

# 6 *Mixed strategies – I. A classification of mechanisms*

One of the clearest predictions of evolutionary game theory is that, in symmetric games, mixed ESS's are often to be expected whenever the potential costs of a contest are large compared to the advantages of winning. Is this expectation borne out? Equally important, what must we know about any particular case before it can reasonably be interpreted as a mixed ESS? These questions are by no means easy to answer, in particular because there are a number of different ways in which the processes of genetic evolution and individual learning can interact in producing stable mixed strategies.

In this chapter, I discuss, with the aid of concrete examples, some of the different ways in which a population may come to show variable behaviour. This discussion leads up to a classification of the mechanisms which underlie such behaviour; the classification is given in Table 10 on p. 78.

The first distinction which must be made is between a mixed strategy, in which the payoffs to the different actions are equal, and a pure strategy of the form 'In situation 1, do *A*; in situation 2, do *B*', in which some individuals are forced to make the best of a bad job. The distinction is best explained by an example. Rohwer (1977) has shown that in winter flocks of the Harris sparrow, individuals vary in the colour of their plumage from dark to pale, and that this variation is correlated with aggressiveness and dominance rank within flocks, the darker birds being more dominant. This could be interpreted as a mixed ESS, following the logic of the Hawk–Dove game. But it could equally well be that birds which, perhaps for environmental reasons, are smaller than average adopt the Dove strategy and develop the corresponding pale plumage at the autumn moult because, if they attempted to be aggressive, they would lose escalated contests. Thus we may be observing a mixed ESS, but equally the birds may be adopting the pure strategy, 'If large, Hawk; if small, Dove'.

I shall return to the specific case of the Harris sparrow in the next chapter, and ask which of these explanations best fits the facts. For the present, the important point is that, whenever one is confronted with a variable pattern of behaviour, one must ask which class of explanation is appropriate: mixed ESS or pure strategy in which animals do different things because they find themselves in different situations? The main criteria for deciding are, first, that in a mixed ESS the different actions should have equal payoffs and, secondly, that the payoffs should be frequency-dependent in the right way, with fitness increasing as frequency decreases.

Perhaps the clearest examples of a pure strategy, with different actions dependent on circumstances, occur when an individual does different things at different ages. For example, in British rivers 75% of male salmon become sexually mature before migrating to the sea (Jones, 1969). They gather round adult mating pairs, and shed competent sperm when spawning occurs. It seems likely that some eggs are fertilised by these precocious males, although there are no estimates of the proportion of eggs so fertilised. This pattern, in which young males attempt to fertilise eggs by adopting tactics different from those they will adopt when adult, is widespread in vertebrates. Typically, this should be seen as a pure strategy: young males adopt the stealing strategy because they could not succeed by adopting the adult one. There are, however, two other possibilities:

(i) The precocious parasitic strategy may be confined to some individuals. Why, in British salmon, are only 75% of males precociously active? The success of precocious activity is likely to decrease as the number of young satellites increases. Therefore, if there is a cost in precocious maturity, there could be a frequency-dependent equilibrium: i.e. an ESS. In the case of salmon this is pure speculation, but on p. 90 I describe an example in which an ESS interpretation is better supported.

(ii) Even if all individuals adopt one strategy when young and a different one when old, we could still be looking at an ESS. Thus imagine a bird species whose members compete – e.g. for territories – in a context such that the payoffs for different actions, say Hawk and Dove, are of a kind to lead to a mixed ESS. One possibility is that the species would become genetically polymorphic. Another is that young birds, aged less than *n* years, should adopt one strategy, and birds *n* years or more in age should adopt the other. Natural selection would act on *n* until the payoffs were equal; the resulting life history strategy would be an ESS.

This bird example is imaginary, but something very like it happens in the evolution of sequential hermaphroditism. The age at which the sex changes is influenced by natural selection (Leigh, Charnov & Warner, 1976). The matter is complicated by the fact that larger size or greater age may be of greater relative benefit to one sex than the other. Ghiselin (1969) suggested that one of the factors favouring the evolution of sequential hermaphroditism is that small animals may perform better in one sex role and large ones in the other. Charnov, Gotshall & Robinson (1978) have pointed out that an absolute difference of this kind is not needed; it is sufficient that one sex should gain *relatively* more than the other by increased size. They use a game-theoretic approach to predict the proportions of the two sexes to be expected in different age or size classes, and test their predictions against data on the shrimp *Pandalus jordani*.

It follows that, even in the apparently simple case of age-related behaviour, it is not always straightforward to distinguish between mixed ESS's and pure strategies. The theoretical distinction is nevertheless quite clear. In a mixed ESS, payoffs of actions depend on the frequencies with which they are performed, and selection has acted to equalise those payoffs. In a pure strategy, individuals choose one action because, as a result of small size, lack of experience etc., they cannot effectively perform any other. Sequential hermaphroditism is best seen as a mixed ESS, because selection has acted to equalise payoffs, but it has done so subject to the constraint that size differentially affects an animal's success in male and female roles.

In view of the unexpected complexity of age-specific behaviour, it may help at this point to give an unequivocal example of a pure strategy in which actions depend on circumstances. Eberhard (1980*a*) describes the behaviour and morphology of the horned beetle *Podischnus agenor*. Males have horns, which are used in fights with other males over the possession of mating sites in sugar canes. As shown in Figure 16, horn size is allometrically related to body size. There are, however, two different allometric relationships, one for small males and one for large males. Males which, for nutritional reasons, are going to be small switch their development onto a different path. These smaller males, with relatively small horns, are

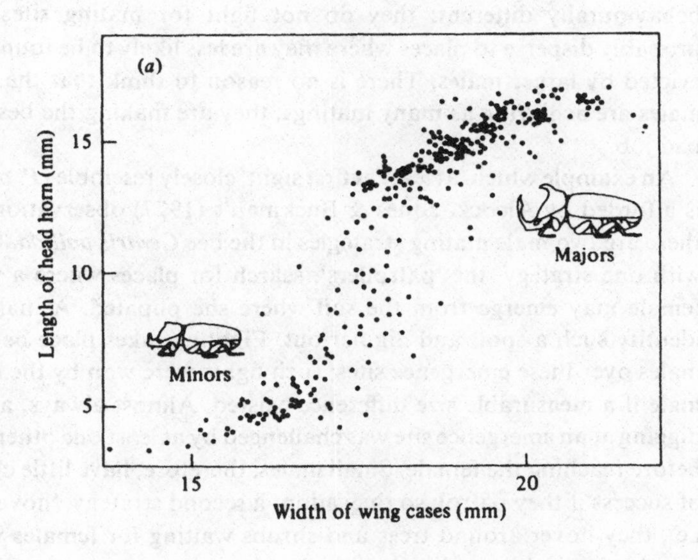


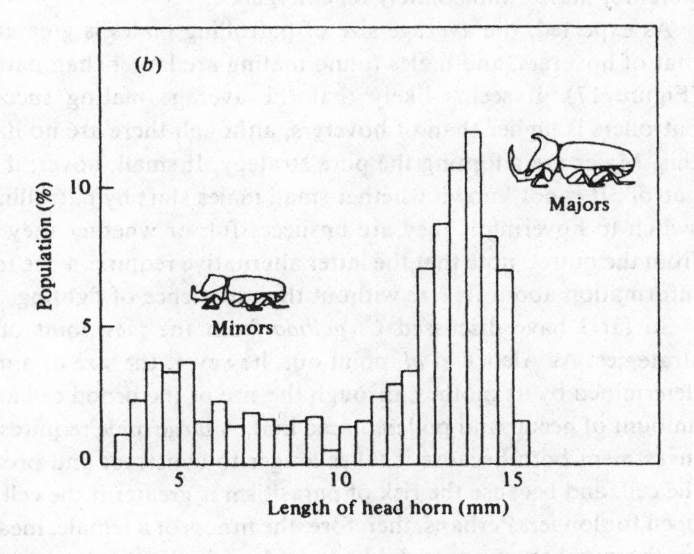


**Figure 16. Allometry in the horned beetle, *Podischnus agenor*. (*a*) Horn size as a function of body size; (*b*) distribution of horn sizes in the population. (After Eberhard, 1980*a*.)**

behaviourally different; they do not fight for mating sites, and probably disperse to places where they are less likely to be found and evicted by larger males. There is no reason to think that the small males are achieving as many matings; they are making the best of a bad job.

An example which, at least at first sight, closely resembles *P. agenor* is afforded by Alcock, Jones & Buckman's (1977) observation that there are two male mating strategies in the bee *Centris pallida*. Males with one strategy, the 'patrollers', search for places where a virgin female may emerge from the soil where she pupated. A male can identify such a spot, and dig her out. Fighting takes place between males over these emergence sites; such fights were won by the larger male if a measurable size difference existed. Almost always, a male digging at an emergence site was challenged by at least one other male before reaching the female. Small males, therefore, have little chance of success if they patrol; so they adopt a second strategy, 'hovering', i.e., they hover around trees and shrubs waiting for females which were not mated immediately on emergence.

As expected, the average size of patrolling males is greater than that of hoverers, and males found mating are larger than patrollers (Figure 17). It seems likely that the average mating success of patrollers is higher than of hoverers, although there are no data on this. Males are adopting the pure strategy 'If small, hover; if large, patrol'. It is not known whether small males start by patrolling and switch to hovering if they are unsuccessful, or whether they hover from the outset; note that the latter alternative requires a bee to have information about its size without the experience of fighting.

So far I have discussed *C. pallida* from the viewpoint of male strategies. As Alcock *et al.* point out, however, the size of a male is determined by its mother, through the size of the brood cell and the amount of nectar and pollen placed in it. A large male requires more investment both because it takes longer to construct and provision the cell, and because the risk of parasitism is greater if the cell is left open for longer. Perhaps, therefore, the fitness of a female, measured by the number of grandchildren she has, is the same whether she produces a few large or many small sons. In support of this interpretation, the variance of size is greater in males than females, so the variability of males cannot merely reflect the inability of females

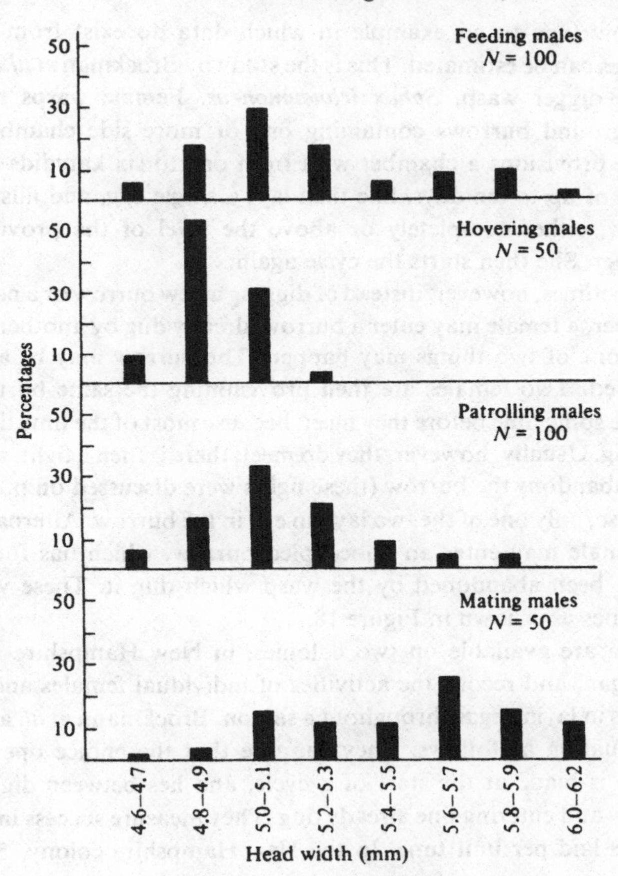


Figure 17. Sizes of different categories of males of the bee, *Centris pallida*. (After Alcock *et al.*, 1977.)

to regulate the quantity of food supplies. Thus the female may be adopting a mixed ESS, and thereby forcing a pure conditional strategy on her sons. There are no data on reproductive success to test this idea. Nor is it known whether individual females specialise in producing either large or small sons, or whether each female is a mixed strategist. This point should be possible to settle, but, if the former were true, it would be hard to discover whether the individual differences are genetic.

I now turn to an example in which data do exist from which fitnesses can be estimated. This is the study by Brockman *et al.* (1979) of the digger wasp, *Sphex ichneumoneus*. Female wasps nest in underground burrows containing one or more side chambers. A female provisions a chamber with from one to six katydids over a period of up to ten days. She then lays a single egg, and fills in the burrow, either completely or above the level of the provisioned chamber. She then starts the cycle again.

Sometimes, however, instead of digging a new burrow or a new side chamber, a female may enter a burrow already dug by another wasp. If so, one of two things may happen. The burrow may be already occupied. Two females are then provisioning the same burrow. It may be some time before they meet, because most of the time they are hunting. Usually, however, they do meet; there is then a fight, and the loser abandons the burrow (these fights were discussed on p. 38). In any case, only one of the two lays an egg in the burrow. Alternatively, the female may enter an unoccupied burrow, which has for some reason been abandoned by the wasp which dug it. These various outcomes are shown in Figure 18.

Data are available on two colonies, in New Hampshire and in Michigan, and record the activities of individual females and their success in laying eggs throughout a season. Brockmann *et al.* analyse the situation as follows. They suppose that the choice open to a female is made at the start of a cycle, and lies between digging a burrow and entering one already dug. They measure success in terms of eggs laid per unit time. In the New Hampshire colony, 59% of

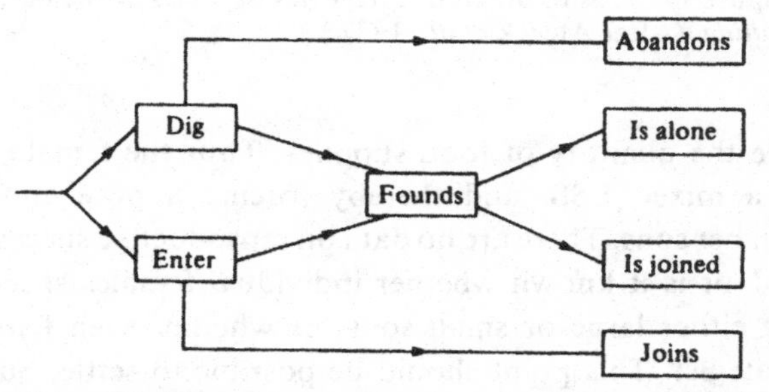


Figure 18. Flow chart for the behaviour of female wasps, *Sphex ichneumoneus*. (After Brockmann *et al.*, 1979.)

choices were to dig and 41% to enter. Entering was a successful strategy if the entered burrow was empty, but an unsuccessful one if it was already occupied. The average success rates, in eggs laid per 100 hours, were 0.96 for digging and 0.84 for entering. These values are close enough to be consistent with the system being at an ESS. Further, it is easy to see that the fitnesses will be frequency-dependent in the required direction; this was confirmed by a computer model.

One puzzling feature is that wasps appear not to distinguish between empty and occupied burrows, although it would pay them to do so. Thus, for those cases in which a wasp entered, the chance of entering an empty burrow was almost equal to the frequency of empty burrows in the colony at the time.

This is not a polymorphism; individual wasps switch from digging to entering and vice versa, and there is no statistically significant tendency for individuals to adopt one or other strategy (Brockmann & Dawkins 1979). Rather surprisingly, the choice is not made on the basis of immediate past success. This raises a serious difficulty. In order to be at an ESS, the probability that each choice will be to enter rather than dig must be close to 0.41. Since the choice appears not to be based on trial and error (perhaps because an individual makes too few trials for learning to be effective), the probability must be a genetic parameter which has been modified by natural selection to the appropriate value, just as a morphological character such as wing length has been modified. The appropriate value, however, must vary from place to place.

In other words, if *S. ichneumoneus* females are adopting a mixed ESS, then the probabilities of entering and digging have been adjusted by natural selection to fit local circumstances. A second population was studied, in Michigan, and found not to be in a stable state. The average success rates from entering and digging decisions were 1.91 and 1.64 eggs per 100 hours, respectively, the difference being statistically significant. The ESS hypothesis can be saved if we suppose that some local populations, because of gene flow or recent environmental change, are some way from the equilibrium. This is entirely plausible; indeed, it is what we would expect. However, the usefulness of the concept of a mixed ESS is lessened if we are obliged to say that populations are as likely to be observed away from the equilibrium as at it.

Three difficulties with the mixed ESS model have been put forward: that animals cannot have a roulette wheel in their heads; that most real contests will be asymmetric; that natural selection will not be strong enough to adjust the probabilities to local conditions. These will be discussed in turn.

*Animals do not have roulette wheels in their heads*

I cannot see the force of this objection. If it were selectively advantageous, a randomising device could surely evolve, either as an entirely neuronal process or by dependence on functionally irrelevant external stimuli. Perhaps the one undoubted example of a mixed ESS is the production of equal numbers of X and Y gametes by the heterogametic sex: if the gonads can do it, why not the brain? Further, in so far as animals can adopt 'probability matching' tactics in experiments on learning, they are demonstrating that they possess the equivalent of a roulette wheel.

*Most real contests are asymmetric*

It was emphasised in Chapter 2 that if some difference exists between two contestants and *if this difference can be perceived by the contestants* then the perceived asymmetry will typically be used as a cue to settle the contest. Most pairwise contests are likely to be asymmetric in this sense. For this reason, we are unlikely to encounter individuals adopting mixed ESS's, or populations at the equivalent genetic polymorphism, when contests are pairwise. The context in which mixed ESS's are likely is that in which the individual is 'playing the field' (p. 23). In fact, the examples I have discussed as candidates for being mixed ESS's (the 1 : 1 sex ratio; age of sex change in sequential hermaphrodites; waiting times in male dung flies; digging or entering in *Sphex*; food provisioning of sons by female *Centris pallida*; dark and pale Harris sparrows) all fall into this category. Of the two examples of pairwise contests so far discussed, one (copulating and intruding male dung flies) is certainly not a mixed ESS, and the other (two female *Sphex* provisioning the same burrow) falls into the rather special category in which the relevant information about asymmetry is only partially known to the contestants.

The examples discussed in the rest of the book tend to confirm the general expectation that mixed ESS's are characteristic of contests against the field, whereas pairwise contests are often settled by asymmetries.

*Probabilities must be adjusted to local conditions*

This difficulty was explained above in the case of *S. ichneumoneus*. There are cases in which it has been shown that contest behaviour does vary from place to place in an appropriate manner. Riechert's (1978) work on the funnel-web spider, *Agelenopsis aperta*, shows this particularly clearly. Fights take place between females over webs. Since, in nature, fights are always between an owner and an intruder, they will be discussed in more detail later (p. 115). For the present, I am concerned only with the way in which the intensity of fighting varies. Riechert studied populations in two habitats, desert grassland and a riparian habitat. Within a habitat, fights lasted longer and involved more dangerous and costly acts when over webs at sites which were known, by independent criteria, to be of greater value. This difference occurred only when fights were between an owner and an intruder, but not if two females unfamiliar with the site were introduced simultaneously. Hence the difference between sites within a habitat depended on the experience of the owner. More immediately relevant, fights in the desert grassland were six times more costly, on average, than in the riparian habitat. This correlates with the fact that good sites are rare, and always occupied, in the desert grassland.

Riechert suggests that the behavioural difference between habitats reflects a genetic difference produced by natural selection. It was shown in the last chapter, however, that a population can reach ESS frequencies in a single generation by learning. In general, learnt and genetic mixed strategies can only be distinguished experimentally. It is important to remember that both learning and genetic change may be involved. Thus, in the relative payoff sum learning rule described in the last chapter, the magnitudes of the 'residuals' determine both initial biases in favour of particular actions, and the rate at which behaviour is modified in the light of experience. These features of learning would be under genetic control. Thus, in the spider example, it is quite likely that there is an initial bias towards aggressive behaviour in the desert grassland population and that some learning occurs.

Table 10. *A classification of mechanisms*

*Difference between strategies genetic*
Gene frequency set by natural selection
I Genetic polymorphism
*Difference between strategies not genetic*
Population genetically homogeneous
A. *Difference between strategies random*
Probability(s) set by natural selection. Populations in different habitats genetically different
IIa Strategy set for life. (Distinguishable from I only by genetic experiment)
IIb Strategy varies from contest to contest
B. *Difference between strategies environmental*
Populations in different habitats can be genetically identical
III Frequency acquired from an environmental cue, but requires a parameter set by natural selection
IV Frequency acquired by trial and error. Developmentally stable strategy, or DSS
V Frequency acquired by cultural inheritance. Culturally stable strategy, or CSS
VI Frequency acquired from an environmental cue, not requiring a parameter set by natural selection. Phenotype fitnesses not equal
*a* asymmetric contests
*b* 'making the best of a bad job'

It is now possible (Table 10) to offer a classification of the mechanisms, genetic and developmental, which can give rise to variable behaviour.

Category I, genetic polymorphism, is clear enough in theory, but can be distinguished from II*a* only by genetic experiment, which is often not practicable. In either I or II*a*, individuals are pure strategists, and selection will tend to equalise their fitnesses.

The distinction between categories II, III and IV needs some explanation, which can best be done with the help of an example. Perrill, Gerhardt & Daniel (1978) studied calling and satellite males in the green tree frog, *Hyla cinerea*. A male may attract females by calling, or may remain silent and attempt to waylay females attracted to a calling male. In 30 field experiments in which a gravid female was released near a pair of males, one calling and the other silent, the former male achieved amplexus on 17, and the latter on 13 occasions.

Some males employed the satellite strategy from night to night, but others changed strategies, sometimes switching during a single night. Clearly, these facts are insufficient to enable us to decide whether fitnesses are equal (particularly because calling may be dangerous), and, if so, how equality is achieved. However, the example is a convenient one to use for explaining some hypothetical mechanisms.

Suppose it turns out that fitnesses are indeed equal. Further, since the ESS frequencies vary from place to place (e.g. with predation risk or frog density), suppose also that the actual frequencies of the two strategies vary correspondingly. How could this come about? I offer three possible mechanisms, corresponding to categories II*b*, III and IV.

Category II*b*: Each frog has a fixed probability $P$ of acting as a satellite, the choice being made at the start of each night, and perhaps again in the course of it. The value of $P$ is the same for frogs in a given place, but varies from one place to another, the causes of these differences being genetic. Thus populations are adapted by natural selection to local conditions.

Category III: Individual frogs adopt the calling strategy if no other frog is calling within a distance $X$; otherwise they act as satellites. Natural selection will modify the value of $X$ until an ESS is reached. The frequency of satellites could, however, vary from place to place in an appropriate way even if all populations were genetically the same.

Category IV: Frogs switch from calling to satellite, or vice versa, according to the success they have achieved with the two strategies in the past. As explained in the last chapter, this could lead to ESS frequencies, and an equalisation of fitnesses, without the need for natural selection to adjust any parameter like $X$ or $P$. It is hard to see how this process could take predation risks into account.

These three mechanisms are not intended to exhaust the possibilities, or even to be particularly plausible; none may be the correct explanation of the behaviour of *H. cinerea*. They do, nevertheless, illustrate three different ways in which a stable mixed strategy, varying appropriately from place to place, might be achieved.

Category V differs from IV in that individuals can transmit what they have learnt to others. To give an extreme example, if each individual learnt what strategy to adopt by copying its mother, this would lead to cultural evolution formally identical to the asexual inheritance which underlies the ESS concept. In practice, cultural evolution will depend on learning from many other members of the population, not from one. Dawkins (1980) has proposed the term 'culturally stable strategy', or CSS, for stable strategies achieved in this way. The idea is pursued further in Chapter 13.

Category VI covers all those cases in which individuals are programmed genetically to adopt a pure strategy in which the particular tactics adopted are conditional on circumstances. In general, the payoffs to different actions will not be equal. It includes all contests in which the choice is influenced by a perceived asymmetry: for example, of size, ownership, age or sex. Such contests are the subject of Chapters 8 to 10. It is, however, important to distinguish two situations. On the one hand, in many species young males adopt tactics different from and less successful than those employed by old males; this is a simple example of category VI. In contrast, consider the hypothetical example on p. 69, in which a bird switched from one behaviour to another at an age which was determined by natural selection so as to equalise the payoffs to the two behaviours. In this case, age is being used, not as an asymmetric cue, but as an alternative to a randomising device in specifying a mixed ESS – category III, if age can be called an environmental cue.

I fear that Table 10 will prove to be neither exhaustive nor unambiguous. It does draw attention, however, to some of the distinctions which should be made when analysing mixed strategies. These will, I hope, become clearer from the discussion of examples in the next chapter.

# 7 *Mixed strategies – II. Examples*

It would be satisfying if one could go through the classification in Table 10 (p. 78), and give examples of each category. Unfortunately, even when it is known that an individual retains the same pure strategy throughout life, there is rarely any direct evidence for genetic involvement; hence it is impossible to distinguish categories I and II*a*. If individuals are known to switch tactics during their lifetimes, and if it can be shown that payoffs for different actions are equal, it may still be difficult to distinguish between II*b*, III and IV. I shall therefore discuss a series of examples of variable behaviour, loosely grouped according to the context in which they occur.

## A The sex ratio

Some of the mechanisms listed in Table 10 can be nicely illustrated by the evolution of the sex ratio. In Chapter 4 section B, it was supposed that the sex ratio is determined by genes acting in one or other parent. This is equivalent to treating it as a game played by females (or by males), each of which is attempting to maximise the number of grandchildren produced. In the simplest case (sons and daughters cost the same; no population structure) the ESS is a 1 : 1 ratio. This can be achieved in two ways. Most commonly, as in birds, mammals and *Drosophila*, there is random segregation of X and Y chromosomes in the meiosis of one or other parent. This is an example of mechanism II*b*. All mothers (or, in birds, fathers) are genetically alike, and have the same probability, 0.5, of producing a son at each conception. It has commonly been supposed that the value of 0.5 is maintained by selection (Fisher, 1930; but see Scudo, 1964, and Maynard Smith, 1980, for a possible alternative view).

In a few animals (e.g. some isopods, cirripedes), however, some females produce only sons and others only daughters. Since the difference between these two classes of females is known to be genetic, this is an example of category I, not II*a*; I know of no case in which two types of female exist, the difference being caused by an environmental cue (II*a*).

In some animals and plants, sex is determined by the environment in which an individual finds itself (Charnov & Bull, 1977). If, as in turtles (Bull, 1980), the environmental cue is incubation temperature, this can still be seen as a game between the parents, since the mother, by choosing the nest site, can choose the sex of her offspring. In other cases, the behaviour of the mother (and hence genes acting in the mother) probably cannot influence the sex of the offspring. If so, we can view the problem as a game between individuals which choose their sex so as to maximise the number of their offspring. For example, in many nematodes which are parasitic in insects during the larval stage and free-living as adults, sex is determined during larval life. Overcrowding leads to small size and maleness, low density to large size and femaleness. This is an example of category III. The 'parameter set by natural selection' is the degree of crowding above which larvae develop as males, which will be selected so as to maximise individual fitness.

Finally, suppose that in our own species a method of choosing the sex of future children is found which is cheap enough to be generally adopted. It seems likely that information will be available about the numbers of boys and girls being born, and that people will select the sex of their children in the light of that information. If so, this would be an example of category IV.

### B Status in flocks

I now return to the case of the Harris sparrow, which was used on p. 68 to explain the difference between a mixed ESS and a pure strategy in which actions are conditional on circumstances. The following account is based on Rohwer (1977), Rohwer & Rohwer (1978) and Rohwer & Ewald (1981). The birds feed during the winter in mixed flocks of varying composition. After an autumn moult, the birds are very variable in colour, between pale and dark; after a second moult, in the spring, all birds are dark. There is a close correlation between plumage and status; in pairwise interactions, the darker bird almost always displaces the paler from a food source.

Free-ranging birds were observed feeding at bait which was distributed in various ways; individuals were also painted, and in some cases given testosterone. Dark birds painted pale continued, as expected, to behave aggressively, and maintained their dominant status, but were involved in many more aggressive interactions in their efforts to maintain dominance. Pale birds painted dark were attacked with greater vigour by naturally dark birds, and in some cases were so persecuted that they were forced to feed away from a flock. When pale birds were simultaneously painted dark and given testosterone, however, their behaviour changed to match their signal, and they successfully maintained a dominant status. Pale birds which were given testosterone but not painted also attacked high ranked birds, but their attacks were disputed and often unsuccessful; these birds eventually dropped out of the flocks.

These results suggest that the population may be at a mixed ESS, with individuals adopting a range of strategies between pale and dark. The alternative view, that pale birds are smaller and weaker individuals which adopt a subordinate strategy because they could not succeed in an aggressive one, is made unlikely by the experiments in which pale birds painted black and given testosterone became successful dominants. If, by the minuscule expenditure of producing more testosterone and more melanin, a bird could become dominant, the advantages of being a dominant cannot be great.

There is some direct evidence for equality of fitnesses. Mean plumage score is similar in early and late winter (Rohwer, personal communication), which suggests that there is little differential mortality. One fact suggests that darker birds may be fitter; there is a tendency for individuals to become darker between their first and second winters but not thereafter (Rohwer, Ewald & Rohwer, 1981). This could be because older birds are more capable of maintaining a dominant position; but there is no necessity for this interpretation. Thus suppose, say, that two-thirds of all birds are in their first winter, and one-third are older, and that the ESS, however maintained, is for one-third of all birds to be dominant and dark. Then the strategy 'be subordinate in the first year and dominant subsequently' would be an ESS, and could not be invaded by a mutant which was always dominant, or one which was always subordinate. Thus the correlation between age and dark plumage *may* indicate that aggression is the more successful strategy, but it need not.

Thus, on balance, the evidence supports this view that the population is at or close to a mixed ESS, and not that pale birds are making the best of a bad job. I would be reluctant to accept this conclusion, however, without some understanding of how the balance is maintained. This question is addressed by Rohwer & Ewald (1981). When seeds are thinly and evenly distributed, there is no trend in the number of seeds eaten per unit time with plumage colour, and only a slight trend in the amount of aggression displayed with plumage colour. When seeds were buried in caches, however, dark birds were much more aggressive than pale ones. Dark birds did not search for caches, but waited until a subordinate found a cache and then displaced it. Aggressive encounters also take place between birds when they are not feeding. These encounters are commonest between birds of similar rank, and are particularly common between pairs of dark birds.

The explanation of these facts seems to be as follows. Dark birds benefit from association with pale birds because they can steal from them, particularly when food is clumped. It pays a dark bird to attack another dark bird and, if possible, drive it away, because a group of subordinate birds is, to a dark bird, a resource worth defending. This aggression between dark birds explains why pale birds painted dark fare so badly. The advantages of aggression clearly decline as aggressive birds become commoner, both because there are fewer subordinates to exploit and because more energy is spent in aggressive encounters. This is enough to produce the required frequency-dependence. One question remains: why do subordinate birds not feed where there are no dominants? One possible answer is that they cannot, because the dominants will seek them out, as suggested by Baker (1978) in a more general discussion of the presence of subordinate birds in flocks. This explanation is made less plausible by the following observations (Rohwer, personal communication). A flock of dominant birds was put in one aviary and a flock of subordinates in another. Running across the back of these two aviaries was a corridor into which strangers of varying rank were placed to make choices. Almost all, regardless of rank, chose to associate with dominants. This suggests that birds gain something by associating with dominants, as implied by Rohwer & Ewald's (1981) analogy of sheep and shepherds; the suggestion is confirmed by the fact that birds in the subordinate flock showed evidence of stress. The nature of the advantage is not clear.

Variable strategies in winter flocks of birds may be widespread. Barnard & Sibly (1981) have shown that individual house sparrows (*Passer domesticus*) are either 'producers' or 'scroungers'. The former spend most of their time searching for food. The latter may also search, but, if searching is expensive, they spend most of their time interacting with others. The commonest form of interaction is to watch searching birds, to move rapidly towards any successful searcher, and to search near to it; less commonly, scroungers follow individual searchers and seize food from them. Individuals persist in one or other strategy, despite changes in the composition of flocks, but it is not clear whether the difference is genetic or acquired. Unlike Harris sparrows, the different strategies are not distinguishable by plumage; also scroungers are, if anything, subordinate, rather than dominant, to producers in aggressive interactions.

It may be rather unusual for individual dominance to be signalled by differences in plumage. Balph & Balph (1979) and Balph, Balph & Romesburg (1979) have studied four flocking species. In two, there is a striking sexual dimorphism in colour. In the evening grosbeak (*Hesperiphona vespertina*), males are more brightly coloured than females and are dominant to them. Flocks are relatively stable in membership, and there are clear intra-sex dominance hierarchies. In Cassin's finch (*Carpodacus cassinii*), older males are reddish, whereas younger males and females are streaked brown and white. Surprisingly, the latter, although smaller, are dominant in aggressive interactions (a nice example of a 'paradoxical' ESS in an asymmetric contest; see p. 102). The dark-eyed junco (*Junco hyemalis*) forms relatively stable winter flocks. There is variability in the darkness of the hood, which is correlated with sex although there is much overlap. Darkness is a predictor of dominance, but only because males are darker than females; within sexes, darkness of hood is not correlated with dominance.

One might attempt a general explanation along the following lines. Sexual differences in plumage have evolved for reasons associated with breeding. Unlike the Harris sparrow, there is no prenuptial moult, so that breeding differences are present also in the winter. Being present, they are used as cues to settle inter-sexual contests. Since winter flocks are relatively stable in membership, individual recognition is possible, so that there has been no need to evolve variable plumage as a rapid indicator of behavioural status.

This explanation fails to account for the overlap in plumage colour between the sexes in the dark-eyed junco. It also runs into difficulties with the fourth species discussed by Balph & Balph (1979), the pine siskin (*Carduelis pinus*). This species forms relatively unstable winter flocks, and has variable plumage (wing stripes varying in width and colour) not closely associated with sex. One would therefore expect the plumage differences to play the same role as in the Harris sparrow, but it seems that they do not. There is some association between dull plumage and dominance. I see no reason *a priori* why bright rather than dull plumage should signal aggressive behaviour in feeding flocks. The puzzle is that the association is relatively weak.

I am left with the strong impression that frequency-dependent selection is generating behavioural variability in feeding flocks of birds. It seems equally clear that the details differ widely between species.

## C Dimorphic males

Gadgil (1972) discussed mechanisms of sexual selection which might lead to a stable genetic polymorphism among males. He argued that the mating success of different investment strategies might depend on the frequency distribution in the population, and that an evolutionary equilibrium would be reached when 'those investing in weapons are just as well off as those which have totally opted out of such investment'. Essentially, Gadgil was describing a mixed ESS. Some examples of male dimorphism of the type he proposed will now be described.

In some species of figwasps, there are two types of males: winged males which disperse before mating, and wingless males with large heads and mandibles which cannot disperse and which fight lethally for the opportunity to mate with females emerging in the same fig. The degree of dimorphism is extreme, but it is not known whether development is switched onto one path or another by genetic or

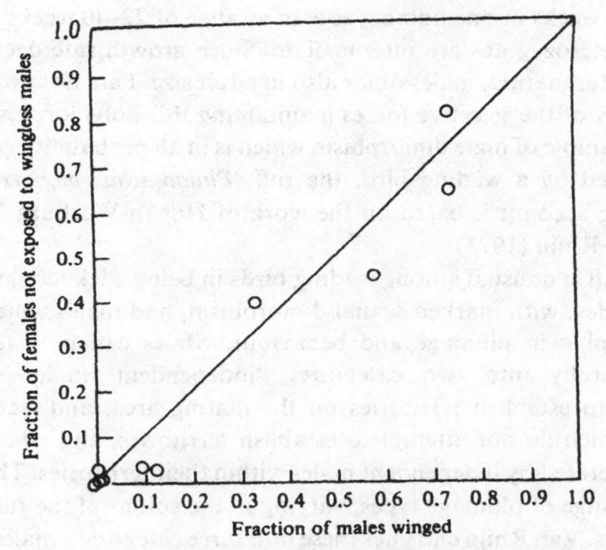


Figure 19. Fraction of migrating female figwasps as a function of the fraction of winged males. (After Hamilton, 1979.)

environmental means. Hamilton (1979) studied 18 species of fig-wasps, in two *Ficus* species, at a site in Brazil. The commonest species tended to have only wingless males, and the rarest only winged males. This makes sense, because in a common species a male is more likely to have females with whom to mate without leaving his natal fig. More interesting are the dimorphic species. As shown in Figure 19, there is a good agreement between the fraction of winged males and the fraction of females leaving their natal fig before mating. Since these females will be mated by winged males, the equality of the two fractions implies equality of fitness; it is also easy to see that the fitnesses will be frequency-dependent in the appropriate way. Thus Hamilton's data provide one of the clearest examples of an evolutionarily stable dimorphism, although it is not known whether it is an example of categories I or II*a* (Table 10, p. 78).

One case in which we know that male polymorphism is genetic was described by Kallman, Schreibman & Borkoski (1973) in the platyfish, *Xiphophorus maculatus*. Two alleles at a sex-linked locus determine the age of maturation of males, which varies from a range of 10–16 weeks in one homozygote to a range of 22–40 weeks in the other; heterozygotes are intermediate. Since growth rate decreases when testes mature, males differ also in adult size. I am not aware of any study of the selective forces maintaining this polymorphism.

An example of male dimorphism which is in all probability genetic is afforded by a wading bird, the ruff, *Philomachus pugnax*. The following account is based on the work of Hogan-Warburg (1966) and Van Rhijn (1973).

The ruff is unusual among wading birds in being a lek (congregating) species, with marked sexual dimorphism, and males which are polymorphic in plumage and behaviour. Males can be classified behaviourally into two categories, 'independent' males which attempt to establish territories on the mating area, and 'satellite' males which do not attempt to establish territories, and which are often tolerated by independent males within their territories. There is a wide range of plumage types, varying in the colour of the ruff and head tufts. Van Rhijn classifies these into three categories: males with 'independent plumage', predominantly dark in colour, which behave as independent males; those with 'satellite plumage', predominantly white, which behave as satellites; those with 'intermediate plumage'. Males with intermediate plumage (Van Rhijn uses the term 'untypical', but this is misleading for a type which accounts for almost half the population) can behave either as satellites or as independents. It is rare for a male to switch from one behaviour to another within a season, although Van Rhijn reports one example of a male which switched temporarily from satellite to (unsuccessful) independent behaviour.

Van Rhijn suggests that the plumage differences are genetic and that the behavioural differences are pleiotropic effects of the same genes. Males with intermediate plumage may be switched into one behaviour or the other by environmental cues. This is plausible, but there is no genetic evidence, and I have been unable to find evidence that plumage patterns remain constant from year to year. Van Rhijn goes further, and suggests that independent and satellite plumages represent the two homozygotes at a locus, birds with intermediate plumages being heterozygotes. This is consistent with the observed frequencies, but would need to be demonstrated experimentally.

The interesting questions are as follows. What maintains the polymorphism? Why do independent males tolerate satellites on their territories? Are the fitnesses of independent and satellite males equal? There are good reasons *a priori* to expect mating success to be frequency-dependent. Satellite males on their own are probably incapable of eliciting a mating response from females. Arenas with few satellite males attract few females; it therefore pays independent males to tolerate satellite males, although it is less clear why it should pay females to go to arenas with many satellites.

Measurements of mating success are sufficient to show that both types of males have appreciable mating success, but not to establish either equality of fitness or the appropriate frequency-dependence. Hogan-Warburg observed that in one locality 20% of satellites achieved 28% of copulations, and in a second locality 39% of satellites achieved only 7% of copulations. These data show strong frequency-dependence of the right kind. Unfortunately, Van Rhijn points out that the two sets of data were collected at different times during the breeding season, and he presents evidence to show that satellite success falls as the season progresses.

In summary, the data on the ruff are entirely consistent with the view that the polymorphism is genetic, and is maintained by frequency-dependent selection with equal phenotypic fitnesses. There are significant gaps in the evidence, on both the genetics and fitness measurements.

Cade (1979) has described an example of alternative male reproductive strategies in the cricket *Gryllus integer*. Males differ markedly in the duration and intensity of calling. These calls attract females, and it seems likely that calling males achieve more matings than silent ones, although quantitative data are lacking. Why, then, should some males remain relatively silent? The answer seems to lie in the existence of a parasitic fly, *Euphesiopteryx ochracea*. Female flies are attracted by the song of the cricket, and deposit larvae on or close to calling males. There was a much higher rate of parasitism in calling males collected in the wild than in non-calling males or in females. Thus there is a trade-off between the mating advantages and the risks of parasitism associated with calling. Males advertising their presence to females may often run the risk of predation.

In the bluegill sunfish, *Lepomis macrochirus*, Gross & Charnov (1980; see also Dominey, 1980) describe the coexistence of two male life history strategies, which they call 'cuckoldry' and 'parental care'. The latter type of male matures at age 7 and constructs a nest; up to 150 nests are arranged in a tightly packed colony. Females are attracted to the nests, where they lay their eggs. The male then fertilises the eggs and provides parental care, without which the young do not survive. The former type of male becomes sexually mature at age 2 and acts as a sneaker, remaining close to the substrate and darting into a nest to fertilise eggs. Later, these males become satellites, resembling females in colouration and behaviour; these males also enter nests and achieve fertilisations. The rate of mortality of these cuckolding males is much higher, and growth is slower; most, and probably all, of them die before they are large enough to become nest-building males.

If this life history dimorphism is to be evolutionarily stable, the expected number of matings achieved by males entering the two pathways must be equal. In the population studied (7 colonies at different depths in the same lake), 21% of males became cuckolders at age 2, while 79% kept growing. In one year, the estimated fraction of all fertilisations achieved by cuckolders was $14 \pm 10\%$. For equal fitnesses, these two estimates should be equal (compare this with the similar argument used by Hamilton below, p. 87). Thus the data are consistent with equal fitnesses; the standard errors, however, are large. M. Gross (personal communication) has found that the fitnesses are frequency-dependent in the appropriate way. It will be interesting to learn the nature, genetic or otherwise, of the developmental switch. Despite these gaps in our knowledge, however, the case is sufficiently remarkable; it is the first recorded case in a vertebrate in which one type of male is wholly parasitic on parental care provided by another.

## D Ideal free distributions

The concept of an ideal free distribution was proposed by Fretwell & Lucas (1970) and further elaborated by Fretwell (1972). In essence, it is a hypothesis about how animals will be distributed in a space composed of habitats of varying suitability, assuming that they are 'ideal' in the sense that each moves so as to maximise its fitness, and 'free' in the sense that each is able to enter any habitat type. The

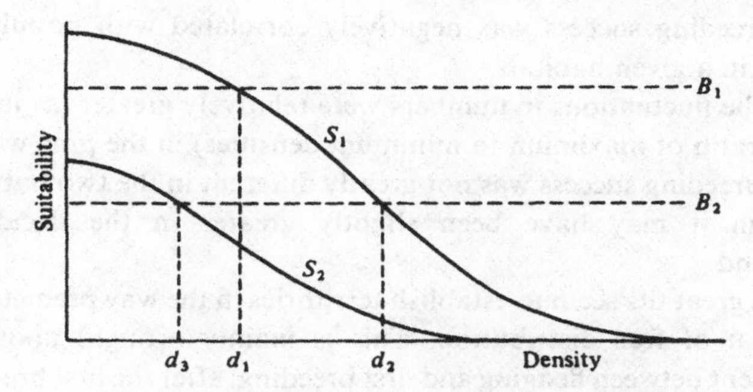


**Figure 20. The ideal free distribution (Fretwell, 1972). $S_1$ and $S_2$ represent the suitabilities of two habitats, as functions of density. For population $B_1$, only habitat 1 will be occupied, at density $d_1$; for population $B_2$, habitats 1 and 2 will be occupied, at densities $d_2$ and $d_3$, respectively.**

conclusion, illustrated in Figure 20, is that animals will so distribute themselves as to equalise the actual fitnesses in different habitats.

Parker's (1970*a*) study of male dung flies, discussed on p. 30, illustrates this principle; males move between cowpats in such a way that, when allowance is made for searching time, mating success is equal on pats of different ages. Parker (1974*a*) shows that the distribution of males around a single pat also conforms to an ideal free distribution. Males may wait on the pat, or in the grass surrounding it, usually upwind of the pat. Arriving females usually fly upwind over the pat, settle on the grass, and walk downwind to the pat; it may be that they avoid landing directly on the pat because, if they do, this results in fights between males which waste the female's time and may put her at risk. The pattern of movement of males between the pat and the grass, as a function of male density and of the age of the pat, are again consistent with the view that fitnesses are equalised.

Kluijver's (1951) data on the breeding success and movements of great tits (*Parus major*) in Holland agree rather well with the predictions of Figure 20. The birds nested in two main habitats, deciduous woods and pine woods; the density of nesting pairs was five to ten times greater in the former habitat. The data agree with the model in the following ways:

(i) Breeding success was negatively correlated with population density in a given habitat.

(ii) The fluctuations in numbers were relatively greater (as judged by the ratio of maximum to minimum densities) in the pine woods.

(iii) Breeding success was not greatly different in the two habitats, although it may have been slightly greater in the deciduous woodland.

Thus great tits seem to establish territories in the way predicted by the ideal of free distribution. This is mainly brought about by movement between fledging and first breeding; after the first breeding season, birds usually return rather precisely to the same spot in subsequent years.

A particular beautiful test of the abilities of animals to distribute themselves in accordance with the ideal free distribution has been described by Witham (1980). The aphid *Pemphigus betae* forms galls on the leaves of cottonwoods. Witham studied the selection of sites by stem mothers. The habitat is a heterogeneous one, both within and between leaves. Witham found that average fitness declines as competition density increases, and that stem mothers adjust their densities in habitats of varying quality in such a way that their average fitnesses are equal in different habitats. He estimates that the average fitness of stem mothers is at least 84% of what it would be if their behaviour was optimal; the true value is higher than this, because in calculating the optimal behaviour it was assumed that females can search for an unlimited time, and no allowance was made for the fact that territorial behaviour in these aphids will interfere with free settlement.

A final example of an ideal free distribution is afforded by Milinsky's (1979) experiments on sticklebacks, which were discussed on p. 63.

## E Dispersal in a uniform environment

The concept of an ideal free distribution concerns the way animals will distribute themselves in a variable environment. Evolutionary game theory has also been applied to the distribution of organisms in a uniform environment (Hamilton & May, 1977). This is a convenient place to discuss that work, although it is not particularly concerned with mixed ESS's.

In their simplest model, Hamilton & May consider a population of parthenogenetic females in an environment consisting of a number of identical sites, each able to support one adult. At intervals, all adults die simultaneously. Each female produces offspring, of which a fraction $m$ disperse to other sites, and $1-m$ remain to compete for the site vacated by their mother. A fraction $p$ of migrants survive, and all non-migrants do so. One individual then becomes an adult at each site, chosen randomly from all offspring at that site.

It is assumed that $m$ is an evolutionary variable determined by the mother's genotype; since mother and offspring are genetically identical, it could equally well be determined by the genotype of the offspring. The evolutionarily stable value of $m$ is shown to be

$$m^* = 1/(2-p). \tag{7.1}$$

Thus even if the probability that a migrant will survive is quite small, it will still pay a female to produce more than 50% migrant offspring. Why should this be so? Qualitatively, it is easy to see that a strategy of producing non-migrant offspring cannot be an ESS. Thus suppose that there are two kinds of female, *A* and *B*; *A* females produce all non-migrant offspring, and *B* females produce some migrant and some non-migrant offspring. A site occupied by a *B* female in one generation can never be occupied by an *A* female in the next, because there are no *A* migrants; in contrast, a site occupied by an *A* female in one generation may be occupied by a *B* female in the next. Hence *A*, producing no migrants, cannot be an ESS.

Hamilton & May go on to analyse more realistic versions of their model, allowing for the fact that females may not all die at the same time, that number of offspring per female may be a random variable (which could be zero), and that the population may be sexual. These changes tend to reduce the evolutionarily stable proportion of migrants, but it remains true that migration will evolve in a uniform habitat. The most interesting modification is the introduction of sex, which introduces an element of parent–offspring conflict (Trivers, 1974). If migrant frequency is determined by genes in the mother, it will be higher than if it is determined by genes in the offspring. Either could be the case: to use their example, a pappus on a seed will be determined by the maternal genotype, and the development of wings by the genotype of the offspring.

# 8 *Asymmetric games – I. Ownership*

The distinction between symmetric and asymmetric games was discussed on p. 22. In this chapter I shall discuss further games which have the following properties:

(i) Every contest is between a pair of individuals one of which is in role *A* (e.g. 'owner', 'larger', 'older') and the other in role *B* (e.g. 'intruder', 'smaller', 'younger').

(ii) Both contestants know for certain which role they occupy.

(iii) The same strategy set (e.g. escalate, retaliate, display, etc.) is available to both contestants.

The role may influence the chances of winning an escalated contest, or the value of winning. More complex contests, in which more than one asymmetry is present, in which there is uncertainty about roles, or in which different strategy sets are available to the two contestants, are discussed in Chapters 9 and 10.

It is convenient to start with the simple Hawk–Dove game, with payoffs shown in Table 11 (which is identical to Table 1, repeated here for convenience). We suppose, however, that each contest is between an owner (e.g. of a territory) and an intruder, and that the contestants know which role they occupy. For the present, we also suppose that ownership does not alter the value of the resource, or the chance of winning an escalated contest. Then, as on p. 22, we can introduce a third strategy, *B* or 'Bourgeois', viz. 'If owner, Hawk; if intruder, Dove'.

We now make the crucial assumption that the probability that an individual occupies a particular role is independent of its strategy. This is equivalent to assuming that there is no pleiotropism or linkage disequilibrium between the genes influencing whether an individual is an owner and the genes influencing its strategy. Of course, for a strategy such as *B*, the role of an individual is correlated with the particular action, Hawk or Dove, adopted; the assumption is that the strategy *B* itself is not correlated with role.

Table 11. *Payoffs for the Hawk–Dove game*

| | $H$ | $D$ |
|---|---|---|
| $H$ | $\frac{1}{2}(V-C)$ | $V$ |
| $D$ | 0 | $\frac{1}{2}V$ |

There is a further possible complication which is ignored when it is assumed that an individual's role is independent of its strategy. The assumption is reasonable if an animal participates in only one contest, or if the outcome of one contest does not influence an animal's role in the next. Things may be more complex if an animal participates in a series of contests. Thus suppose the contests are for territories, or some other resource which lasts for an appreciable time. Then an animal adopting Dove in one contest is more likely than a Hawk to be a non-owner in the next contest. If so, strategies are not independent of roles. This effect is ignored in what follows, but will have to be allowed for when analysing certain types of field data.

Given independence between roles and strategies, the payoff matrix is shown in Table 12. In deriving the matrix, note that:

(*a*) The $B$ strategist chooses $H$ and $D$ with equal frequencies, because it is an owner on half the occasions and an intruder on half.

(*b*) In contests between two $B$'s, if one chooses $H$ then the other chooses $D$.

If $V > C$, the only ESS is $H$; if it is worth risking injury to gain the resource, ownership will be ignored. The more interesting case is $V < C$. Then the only ESS is $B$; ownership settles the contest without escalation.

Before giving examples of this principle, two points are worth noting:

(i) The conclusion does *not* depend on the value of the resource, or the chance of victory, being different for owner and intruder. Maynard Smith & Parker (1976) used the phrase 'uncorrelated asymmetry' to refer to such contests. This seems to have led to some confusion. People have supposed that we were arguing that, in real

Table 12. *The Hawk–Dove–Bourgeois game*

| | $H$ | $D$ | $B$ |
|---|---|---|---|
| $H$ | $\frac{1}{2}(V-C)$ | $V$ | $\frac{3}{4}V-\frac{1}{4}C$ |
| $D$ | 0 | $\frac{1}{2}V$ | $\frac{1}{4}V$ |
| $B$ | $\frac{1}{4}(V-C)$ | $\frac{3}{4}V$ | $\frac{1}{2}V$ |

cases, ownership does not alter payoffs. We did not argue for this view, and do not hold it. Our point was that there is no *need* for differences in payoff before asymmetries will settle contests. Consequently it is a logical error, rather than a factual one, to conclude that because, in some particular case, ownership does settle contests, there must therefore be a payoff or resource-holding power difference associated with ownership. There may or may not be such a difference, but its presence cannot be deduced from the fact that ownership settles contests.

(ii) What has been said of the strategy $B$ applies equally well to the strategy $X$, 'if owner, $D$; if intruder, $H$'. The difficulty with strategy $X$ is that it leads to a kind of infinite regress, because as soon as an owner loses a contest it becomes an intruder, and hence able to win its next contest. We are unlikely, therefore, to meet strategy $X$ when the contested resource is of value only if it can be held for a long time. As Parker (1974*b*) suggested, however, if the resource is a feeding or drinking station, the strategy 'if the station has been held for more than time $T$, then $D$; otherwise $H$' might well evolve. The only case known to me of strategy $X$ being adopted when the resource is of more permanent value is the account by Burgess (1976) of the social spider *Oecibus civitas*. These spiders live in groups, but each constructs its own web and refuge hole. If a spider is driven from its hole, it may dart away and enter the refuge of another spider of the same species. Then, to quote Burgess, 'If the other spider is in residence when the intruder enters, it does not attack, but darts out and seeks a new refuge of its own. Thus once the first spider is disturbed the process of sequential displacement from web to web may continue for several seconds, often causing a majority of spiders in the aggregation to shift from their home refuge to an alien one.'

Despite this fascinating and curious example, it is far commoner for contests to be settled in favour of owners. To establish that ownership is the cue which settles a particular type of contest, it is not sufficient to show that, in the field, contests are usually won by owners. It might be that contests are settled in accordance with size or strength, either by assessment or actual combat; if so, owners, being past winners, are likely to be stronger and hence to win more contests even if ownership itself is not used as a cue. It is, therefore, desirable to establish two further points; first, each of two animals will win a contest, according to which is the owner, and secondly, an escalated contest will ensue if each of two contestants perceives itself as the owner. Some examples in which these additional points have been established will now be described. In each case there are complications; not surprisingly, the *H–B–D* model is only a partial picture of what is happening.

Kalmus (1941) trained Italian bees to feed on sugar at a training table. So long as there was plenty of sugar, the bees tolerated foraging Caucasian bees. If food was short, however, Caucasian bees were driven off. In a second experiment, Caucasian bees were trained to feed at the table; they then drove off foraging Italian bees when food was scarce. Both types could be trained to the same food source by supplying copious food. If food then became scarce, Kalmus reported that 'a general battle ensues'. Formally, this meets the requirements mentioned above needed to show that ownership is the relevant cue. However, I have two reservations. First, it may be that the cue determining whether a bee will fight to defend a food source is not prior ownership but the number of other workers present from its own hive. Secondly, Kalmus found that active defence of a food source was not made against bees from a different hive but of the same strain, although the hive is defended against such workers; hence the behaviour may not be very relevant in the field.

A more complete demonstration is afforded by Kummer, Götz & Angst's (1974) study of contests between male Hamadryas baboons over females. In the wild, a male Hamadryas forms a long-lasting association with several females. It was shown that if male *A* was permitted to form a bond with a strange female, then a second male, *B*, who has watched the interaction will not subsequently challenge *A* for ownership. If, on a later occasion, male *B* forms a bond with a female, he will not subsequently be challenged by $A$. Escalated fights do occur between two males if each perceives himself as the owner of the same female. It seems clear that ownership, and not any perceived difference in size or strength, is decisive in settling contests. Bachmann & Kummer (1980), however, have shown that female choice can have some influence on the outcome. In an experimental situation, low- and middle-ranking males showed greater respect for an owning male if the female preferred the owner in choice tests; dominant non-owning males did not alter their behaviour in response to female preferences. This makes sense, because if a female prefers a male she is more likely to stay with him, and is therefore more valuable to him. There is some evidence that female choice is relevant in the wild. Abegglen (1976) observed a troop in which male fighting had resulted in extensive redistribution of females. Several mother–daughter pairs were separated by the fighting, but were found to be reunited months later, indicating that their preferences had influenced the course of events.

To summarise on Hamadryas baboons, there are escalated fights between adult males over females, and there is evidence that female choice can affect both the initiation and outcome of such fights. Nevertheless, most potential contests are settled by prior ownership, and need not depend on perception by the contestants of differences in fighting ability.

Davies (1978) studied territorial behaviour in the speckled wood butterfly, *Pararge aegeria*. Males defend patches of sunlight on the floor of the woodland, moving as the sun moves. In Davies' study, some 60% of males held sunspots at any one time; the remainder patrolled in the canopy. If the sun was not out, all males patrolled lower down. When another insect flies past, the territorial male flies up to meet it. If the intruder is a female, the pair usually settle and courtship ensues; if the female is a virgin, they then fly up into the canopy, where mating can sometimes be seen to take place. If the intruder is a male, a brief spiral flight (3–4 seconds) ensues, after which one male returns to the territory and the other to the canopy. By marking individuals, it was found that it was always the owner who returned to the territory. Finally, if the intruder is of a different species, the territorial male ignores it and returns to the territory.

If a male was removed from a territory, it was replaced in a few minutes by a male flying down from the canopy. Males do not feed in their territories, but those in patches of sunlight meet and court more females. Male survival was approximately exponential, with a mean of 7 days. Males marked in the canopy usually obtained a territory later, sometimes on the same day, if new territories became available as the sun moved or emerged from behind clouds. Hence, to hold a territory does increase fitness, but not very greatly so.

Davies was able to show that it is ownership which decides contests, the 'owner' being a male which has settled in a territory, if only for a few seconds. If a male was removed from a territory, held in a net until a new male had settled, and then released, it was in all cases the new owner which retained the territory. Surprisingly, the contests on these occasions did not last any longer (3–4 seconds) than typical ones. It was possible to allow each of a pair of males to own the same territory in turn, and show that each successfully defended the territory against the other. Finally, Davies was able on a few occasions to introduce a second male into a territory without the owner noticing. Then, when one or other male flew up it was immediately challenged by the other, and a spiral flight ensued which lasted on average for 40 seconds, ten times longer than typical ones.

As in Kummer's study of Hamadryas baboons, these results fit beautifully with the predictions of the Hawk–Dove–Bourgeois model, but again there are complications. The main one is that if there is a large patch of sunlight, it is occupied by several males which tolerate one another's presence. Davies found that the number of males per patch fits the ideal free distribution (p. 90): the number of encounters with females each male can expect is independent of patch size. It is not clear, however, how this is brought about.

Similar results were obtained by Dr L. E. Gilbert (personal communication) on male swallowtails, *Papilio zelicaon*, which hold hilltop territories. Owners always win, and there was an escalated contest if two males perceived themselves as owners (they had been permitted to own the same hilltop on alternate days). Baker (1972) found a more complex situation in the peacock, *Inachis io*. Males hold patches of nettles as territories; these are oviposition sites for females. Typically, contests are won by the owner after a spiral flight. Sometimes, however, a second male may settle in a territory, perhaps because the owner is away courting a female. A longer spiral flight then ensues; this is usually but not always won by the original owner, probably because the intruder cannot find its way back to the territory when the spiral flight is broken off. If both males do return to the territory, a succession of spiral flights takes place, the eventual winner being the stronger flier, which is able to keep above and behind its opponent.

One final example of the significance of ownership in settling contests is taken from the work of Drs A. Pusey and C. Packer on lions (personal communication). Groups of males co-operate to take over female prides, but once in control of a pride, males compete for oestrous females. A male forms an exclusive consortship with an oestrous female and prevents other males from coming into close proximity to the female. As long as the consorting male has clear ownership of the female a rival male will not seriously challenge the owner for the female. One male may be owner of the female during one oestrous period but be a rival for the female during the next. A male can consort continuously with a female throughout her oestrus, which lasts for several days. There is, however, competition between males to be the first to establish ownership; in particular, a male may guard a female for several days before she shows signs of receptivity.

Particular interest attaches to those situations in which two males in the same cooperating group fight over possession of females. These fights occur in two situations, which have in common that the asymmetry between owner and non-owner has broken down. The first, and more obvious, case arises when the owner wanders too far from the female, enabling an intruder to come closer to her; ownership is then unclear. The second case arises when two consort pairs come into close proximity. There is then no longer an asymmetry, and a fight may ensue; one male may try to acquire the other's female and thus may come to control two females simultaneously, but in some cases no such attempt is made, and the fight seems to result merely from the intolerance felt by an owner of the presence of a second male. This latter case affords a dramatic (because counter-intuitive) example of the importance of asymmetries in settling contests. In thinking about it, it is important to remember that the cost of a contest between male lions is high. Not only is there risk of injury in the contest itself; even an uninjured male would pay a price if its opponent was injured, because a group of

Table 13. *The Hawk–Dove–Bourgeois game when the value of the resource is V to the owner and v to the intruder*

| | $H$ | $D$ | $B$ | $X$ |
|---|---|---|---|---|
| $H$ | $\frac{1}{4}(V+v-2C)$ | $\frac{1}{2}(V+v)$ | $\frac{1}{4}(2V+v-C)$ | $\frac{1}{4}(V+2v-C)$ |
| $D$ | 0 | $\frac{1}{4}(V+v)$ | $V/4$ | $v/4$ |
| $B$ | $\frac{1}{4}(V-C)$ | $\frac{1}{4}(2V+v)$ | $V/2$ | $\frac{1}{4}(V+v-C)$ |
| $X$ | $\frac{1}{4}(v-C)$ | $\frac{1}{4}(V+2v)$ | $\frac{1}{4}(V+v-C)$ | $v/2$ |

males in which some are injured is less likely to be able to defend the female pride against other groups. Because the price is high, dependence on the asymmetry will be strong, and the risk of escalation on the relatively rare occasions when the asymmetry breaks down correspondingly great.

The assumption that the payoffs to owner and intruder are equal, which is made in deriving Table 12 (p. 96), is often unrealistic. Thus the value of a territory may be greater to an owner, who has already learnt about the distribution of food, refuges, etc. In some cases ownership may confer advantages in an escalated contest. Inequality of payoffs is also likely for other types of asymmetry, for example, of size or age. Inequality of payoffs is not necessary if an asymmetry is to settle contests (although this conclusion will be modified later, when discussing the asymmetric war of attrition). If, however, inequalities of payoff do exist, they can influence outcomes.

I first discuss the asymmetric Hawk–Dove game; in doing so, the 'infinite regress' problem mentioned above (p. 96) will be ignored, because I want the discussion to apply to asymmetric contests in general, not only to contests involving ownership. I then turn to the conceptually more difficult problem of the asymmetric war of attrition.

Consider the contest described on p. 96, but suppose that the value of the resource is $V$ to the owner and $v$ to the intruder; qualitatively similar conclusions would follow if we supposed that the value of the resource was the same for owner and intruder, but that their chances of winning an escalated contest were different. There are four possible strategies: $H$, Hawk; $D$, Dove; $B$, adopt $H$ if owner and $D$ if intruder; $X$, adopt $D$ if owner and $H$ if intruder. The payoff matrix is shown in Table 13. It is easier to see what will happen if we consider two

Table 14. *Two numerical examples of the game shown in Table 13: case (i) $V=16$; $v=8$; $C=20$; case (ii) $V=20$; $v=8$; $C=16$*

| | Case (i) | | | | | Case (ii) | | | |
|---|---|---|---|---|---|---|---|---|---|
| | *H* | *D* | *B* | *X* | | *H* | *D* | *B* | *X* |
| *H* | −4 | 12 | 5 | 3 | *H* | −3 | 14 | 8 | 5 |
| *D* | 0 | 6 | 4 | 2 | *D* | 0 | 7 | 5 | 2 |
| *B* | −2 | 11 | 8 | 1 | *B* | 2 | 13.5 | 10 | 3 |
| *X* | −6 | 7 | 1 | 4 | *X* | −4 | 7.5 | 3 | 4 |

numerical cases (Table 14). Case (i) has two pure ESS's, namely *B* and *X*. *B*, which escalates when the value of victory is higher, can be called the 'common-sense' ESS, and *X*, which escalates when the value is lower, a 'paradoxical' ESS. In case (ii), only the common-sense ESS exists. From Table 13, a paradoxical ESS exists only if $v/2 > \frac{1}{4}(V+2v-C)$, or $V<C$; otherwise *H* can invade.

So far we have shown only that *B*, and sometimes *X*, are stable against other pure strategies. Are they stable against invasion by a mixed strategy? If neither *B* nor *X* were possible strategies, the only ESS would be a mixed one, adopting *H* and *D* with probabilities (0.6, 0.4) in case (i) and (0.7, 0.3) in case (ii). Calling this mixed strategy *M*, we can draw up the payoff matrices in Table 15.

Nothing new emerges in case (ii); *B* is still the only ESS. For case (i), however, it turns out that although the paradoxical strategy, *X*, is an ESS, it could not invade a population adopting the mixed strategy. Thus, if we start from a population of animals which ignore the asymmetry, and therefore adopt the mixed strategy *M* (or, perhaps, reach the corresponding genetic polymorphism), a *B* mutant could invade, leading to the evolution of the common-sense ESS, but there is no way in which the paradoxical ESS could be reached, even if it is stable.

It seems therefore, that paradoxical ESS's are a mathematical curiosity which will rather rarely be realised in nature. Suppose, however, that a species is at a common-sense ESS for an asymmetric game, and that the payoffs then change so as to render that ESS paradoxical and the reverse strategy a common-sense one. Provided

Table 15. *Payoff matrices for the game shown in Table 14 when a mixed strategy, M, is possible*

| Case (i) | | | | Case (ii) | | | |
|---|---|---|---|---|---|---|---|
| | *M* | *B* | *X* | | *M* | *B* | *X* |
| *M* | 4 | 4.5 | 2.5 | *M* | 2.1 | 7.1 | 4.1 |
| *B* | 4.5 | 8 | 1 | *B* | 5.45 | 10 | 3 |
| *X* | 0.5 | 1 | 4 | *X* | −0.55 | 3 | 4 |

that the payoffs do not change so drastically as to render the paradoxical ESS unstable, the population will retain that strategy, even though it could not have acquired it if payoffs had remained constant. I know of no example in which contests are won conventionally by the contestant to whom the value of winning would be lower. There are, however, examples of the mathematically similar situation, in which contests are won conventionally by the contestant least likely to win an escalated contest. One case, of dominance in winter flocks of Cassin's finch, was mentioned on p. 85.

These examples of paradoxical ESS's should perhaps not be taken too seriously. Such ESS's will be rare, because they cannot easily evolve from a population ignoring the asymmetry and hence adopting a mixed ESS. In the symmetric war of attrition, for a resource of value $V$, the ESS is to be prepared to expend an amount $x$, where $p(x)=\exp(-x/V)/V$. Suppose now that every contest is between an owner and an intruder, but that ownership does not alter $V$. In an earlier discussion (Maynard Smith, 1974), I wrongly claimed that the new ESS is 'choose $M$ when owner; choose 0 when intruder', where $M>V$. Call this strategy $B$, and the strategy which ignores ownership $I$. The payoff matrix is shown in Table 16.

Table 16. *Payoff matrix for the asymmetric war of attrition*

| | *I* | *B* |
|---|---|---|
| *I* | 0 | $Ve^{-M/V}$ |
| *B* | 0 | $V/2$ |

Table 17. *Contest types in which an individual, 'ego', is engaged, in the asymmetric war of attrition*

| | Ego's choice | Opponent's choice | Frequency |
|---|---|---|---|
| Ego is owner Value $V$ | $M$ | 0 | $(1-F)^2/2$ |
| | $M$ | $M$ | $F(1-F)/2$ |
| | 0 | 0 | $F(1-F)/2$ |
| | 0 | $M$ | $F^2/2$ |
| Ego is intruder Value $v$ | 0 | $M$ | $(1-F)^2/2$ |
| | 0 | 0 | $F(1-F)/2$ |
| | $M$ | $M$ | $F(1-F)/2$ |
| | $M$ | 0 | $F^2/2$ |

If $M > V$, then $E(I,B) < V/e < E(B,B)$, so $B$ cannot be invaded by $I$. Following this line of argument, Maynard Smith & Parker (1976) considered contests in which the value $V$ to the owner is greater than the value $v$ to the intruder, and concluded (as in the Hawk–Dove game just analysed) that both common-sense and paradoxical ESS's exist, but that only the common-sense strategy can invade a population adopting $I$.

The trouble with this argument is that it does not explain how $M$ is to be fixed. Since in a pure $B$ (or pure $X$) population, any positive value of $M$ is as good as any other, the value will drift.

The natural way out of this difficulty, suggested by Parker & Rubinstein (1981) and analysed further by Hammerstein & Parker (1981), is to suppose that, with a low frequency $F$, individuals mistake their roles. Thus imagine a population adopting the strategy $B$, 'choose $M$ when owner; choose 0 when intruder', where $M$ can be a pure or mixed strategy. Table 17 shows the possible kinds of contest, and their frequencies, in which an individual, 'ego' is engaged.

The only contests relevant to the evolution of $M$ are those in which both contestants choose $M$. For ego, as for his opponent, these occur with equal frequency when the reward for victory is $V$ or $v$. Hence, at the ESS, $M$ is given by the probability distribution,

$$M = p(x) = (1/\bar{V})\,e^{-x/\bar{V}}, \tag{8.1}$$

where $\bar{V} = (V+v)/2$.

We now ask whether a mutant playing other than 0 when intruder could spread. The answer is that it could not, provided $v < V$. It follows that the common-sense strategy 'choose $p(x)$ if owner; choose 0 if intruder' is an ESS, and the paradoxical strategy is not an ESS, provided $v < V$. Further, the asymmetric war of attrition with $v = V$ does not have a stable Bourgeois ESS, because such a strategy would be invaded by selectively neutral mutations which ignored the asymmetry.

To summarise, asymmetric games of the Hawk–Dove type, with a finite set of discrete possible pure strategies, can have both common-sense and paradoxical ESS's, but only a mutant adopting the former strategy can invade a population whose members ignore the asymmetry and adopt the appropriate mixed strategy. Asymmetric games of the war of attrition type, with a continuously distributed set of possible actions, can be analysed if one assumes that errors in role identification occur. If so, and if payoffs in the two roles are unequal, only the common-sense ESS exists; that is, the contestant to whom the value of winning is greater wins, and the other contestant gives in cheaply.

In the asymmetric war of attrition with equal payoffs a mathematically pathological situation arises, with a set of equivalent, neutrally stable, equilibria, varying from a strategy which adopts $p(x) = (1/V) \exp(-x/V)$ regardless of role to one which adopts $p(x)$ when owner and 0 when intruder. In practice, the population would be tipped towards the common-sense ESS, either by some inequality in payoff or chances of winning, or by lack of continuity in the set of possible plays.

A final question concerns the relative suitabilities of the Hawk–Dove and war of attrition models – the comparison arising because the former model indicates the existence of types of ESS, of which paradoxical ESS's are an example, not permitted by the latter. The essential difference is between a discrete set of possible actions and a continuous one. Thus suppose that, in passing from display to physical contact, an animal puts itself in a position from which it cannot escape without some finite and uncontrolled risk of injury: a Hawk–Dove model would then be appropriate. In contrast, a contest which can be broken off at any time, without risk, is better treated as a war of attrition.

# 9 *Asymmetric games – II. A classification, and some illustrative examples*

In the last chapter, I discussed contests having a single asymmetry which is unambiguously known to both contestants at the start of the contest; the obvious example is the asymmetry between the owner of a resource and an intruder. A number of complexities arise in actual animal contests. The following classification is not exhaustive: it is intended mainly as a guide to Chapters 8–10.

1. A single asymmetry initially present, known certainly to both contestants.
   (*a*) Asymmetry uncorrelated with payoff or RHP ('resource-holding power'.
   (*b*) Payoff and/or RHP differ between the two roles.
   (*c*) Strategy sets as well as payoffs and RHP differ between the two roles (e.g. male–female and parent–offspring contests).

Types 1(*a*) and 1(*b*) were the topic of the last chapter. Type 1(*c*) is discussed in Chapter 10.

2. A single asymmetry present, but each contestant knows only its own state. This is the 'game with random rewards', which was discussed in Chapter 3, and Appendix G.
3. A single asymmetry present, but information about it is uncertain (e.g. differences in size or strength). Such contests involve a phase of 'assessment'. Considerable theoretical difficulties arise if the information acquired during assessment is uncertain.
4. More than one asymmetry present.

This chapter deals with types 3 and 4. First, I discuss the case in which unambiguous information about the asymmetry can be acquired. This raises no particular difficulties; as expected, an assessed asymmetry can settle contests without escalation. Some illustrative examples are then discussed. I then turn to the more difficult, but perhaps more realistic, case in which asymmetries of size and ownership are simultaneously present. Finally, using the spider *Agelenopsis* as an illustrative example, I discuss the case in which there are also differences in the value of the resource, but these differences are known only to the owner.

First, however, a theorem of Selten (1980) will be discussed. This states that a game which has an asymmetry known with certainty to both contestants cannot have a mixed ESS; that is, no mixed strategy can satisfy conditions (2.4*a*, *b*). To see why this is so, we proceed by a *reductio ad absurdum*. Thus imagine that a contestant can be in role 1 or role 2, and assume that a mixed ESS exists: 'In role 1, choose *I*; in role 2, choose *J*'; where *I* is the *mixed* strategy, 'choose *A* with probability *p* and *B* with probability $1-p$'; *J* can be pure or mixed. Then, by the Bishop–Cannings theorem (Appendix C), the payoffs to *A*, *B* and *I* against *J* must be equal. Therefore, to show that *I* is an ESS, we would have to show that *I* does better against *A* than *A* does against itself, and similarly for *B*. But we cannot do this, because *I*, *A* and *B* are all appropriate to role 1, and so never meet one another. In other words, there is no way in which our supposed ESS can satisfy the criteria (2.4*a*, *b*).

Selten's theorem can be proved more rigorously. We must be clear, however, about what has been proved. This is that no mixed strategy can satisfy conditions (2.4*a*, *b*). However, it *is* possible for a mixed strategy *I* to be *neutrally* stable, in the sense that *I*, *A* and *B* are all equally good. In fact, we have already met such a case when discussing the asymmetric war of attrition (p. 105). Thus suppose a resource is worth *V* to the owner and *v* to an intruder, where $V > v$. Then, in a war of attrition, the strategy 'choose $p(x) = \exp(-x/V)/V$ when owner; choose 0 when intruder' is a mixed one, but it is not an ESS, because it is only neutrally stable against mutants choosing any other cost when owner. To analyse the game, we had to suppose that errors in role identification occur, so that intruders do sometimes choose $p(x)$. Then Selten's theorem no longer holds, because roles are not certainly known; it can be shown that now $p(x)$ is indeed an ESS. Thus Selten's theorem cannot rule out the possibility that an asymmetric game has a neutrally stable mixed strategy, which becomes an ESS if errors of role identification occur.

With Selten's theorem in mind, we can now consider the role of 'assessment' in animal contests. Suppose first that both contestants

Table 18. *The Hawk–Dove–Assessor game*

| | $H$ | $D$ | $A$ |
|---|---|---|---|
| $H$ | $\frac{1}{2}(V-C)$ | $V$ | $\frac{1}{2}(V-C)$ |
| $D$ | 0 | $\frac{1}{2}V$ | $\frac{1}{4}V$ |
| $A$ | $\frac{1}{2}V$ | $\frac{3}{4}V$ | $\frac{1}{2}V$ |

can unambiguously distinguish some difference, for example in size, which is a perfect predictor of which would win an escalated contest. We can now introduce a new strategy into the Hawk–Dove game, $A$, or 'Assessor', which chooses Hawk if larger and Dove if smaller. The payoff matrix is shown in Table 18. As expected, if there is a cost to escalation, $A$ is the only ESS of the game.

We can now complicate the game in two ways. Suppose that the assessment phase itself costs both participants $c$, where $c < C$, the cost of losing an escalated fight. Suppose also that although the 'size' difference can be estimated unambiguously, it is not a perfect predictor of which contestant would win an escalated fight. Thus an animal may be certain it is the larger, but not that it would win a fight. Let $x$ be the probability that the larger animal would win. The payoff matrix is shown in Table 19. Note that in this and later payoff matrices for asymmetric games, the lower entry in each box is the payoff to the contestant adopting the strategy on the left, and the upper entry to the contestant adopting the strategy above. In seeking for the ESS of this game, we know from Selten's theorem that only pure ESS's can exist; note that if the assessment process led to ambiguous results, so that an animal was uncertain whether it was the larger, the theorem would no longer apply.

Assessor is an ESS if

(i) $c < \frac{1}{2}V$

and (ii) $Cx > V(1-x)$.

Hawk is an ESS if

(i) $c < \frac{1}{2}(V-C)$

and (ii) $Cx < V(1-x)$.

Note that Assessor and Hawk can never be alternative ESS's for the

Table 19. *The Hawk–Dove–Assessor game when there is a cost c of assessment and the larger contestant has a probability x of winning an escalated contest. In each box, the lower entry is the payoff to the strategy on the left*

| | *H* | *D* | *A* |
|---|---|---|---|
| *H* | $\frac{1}{2}(V-C)-c$ / $\frac{1}{2}(V-C)-c$ | $0$ / $V$ | $\frac{1}{2}[Vx-C(1-x)]-c$ / $\frac{1}{2}[V(1-x)-Cx]+\frac{V}{2}-c$ |
| *D* | $V$ / $0$ | $V/2$ / $V/2$ | $V$ / $0$ |
| *A* | $\frac{1}{2}[V(1-x)-Cx]+\frac{V}{2}-c$ / $\frac{1}{2}[Vx-C(1-x)]-c$ | $0$ / $V$ | $\frac{1}{2}V-c$ / $\frac{1}{2}V-c$ |

same parameter values: after assessment, it either is or is not worth while for the smaller contestant to continue.

Factors favouring Assessor as an ESS are:

(i) Assessment is cheap ($c$ is small).
(ii) Escalation is dangerous ($C$ is large).
(iii) Size is a good predictor of victory ($x \simeq 1$).

The third factor is by no means essential. Indeed, if escalation is sufficiently dangerous ($C \gg V$), Assessor can be an ESS even if $x < 0.5$. This is an example of a paradoxical strategy (see p. 85); 'escalate if smaller, retreat if larger'.

Maynard Smith & Parker (1976) extended the analysis of a similar model to cases in which size assessment is itself uncertain. This introduces the following difficulty. Even in a population consisting wholly of 'Assessors', escalated contests occur when the smaller of two contestants mistakenly estimates that it is the larger. Despite this difficulty, it turns out that the Assessor strategy, 'escalate if estimate opponent is smaller, display if estimate opponent is larger', is stable for a wide range of parameter values. In practice, the main importance of inaccurate assessment may be that, even when a population is at an 'Assessor' ESS, escalated contests will still occur, although with low frequency.

Are assessment strategies observed in nature? What we need to show is the following:

(i) A difference in some property of the contestants is perceived by them, and serves to settle contests without escalation.

(ii) Behaviour during the first phase of a contest enables animals to perceive the difference.

(iii) The property should be such that it is expensive in resources to signal a high value; otherwise, assessment strategies would be vulnerable to cheating.

(iv) The property should be correlated with fighting success. As already explained, this condition is not essential, but it is to be expected.

A number of authors have pointed out that organs used in fighting are displayed prior to fighting, and that such displays may settle contests without escalation. Geist (1966) has shown that, in Stone's sheep (*Ovis dalli stonei*), horn size is more variable than body size; that horns are displayed in agonistic encounters between the males; and that, when a ram enters a new band, the great majority of its interactions are with sheep with the same degree of horn development. It seems likely that assessment of horn size is used in determining, without escalation, the position of the new ram in the dominance hierarchy. In similar vein, Packer (1977*a*) has pointed out that fighting ability of baboons (*Papio anubis*) declines with wear and injury to canines, and that the canines are shown in yawning displays during agonistic encounters. Morton (1977), expanding on an earlier remark by Collias (1960), pointed out that, in both birds and mammals, low-pitched sounds are usually associated with aggression and high-pitched ones with fear and appeasement. He suggests that this association has evolved because low pitch is usually associated with large size. More detailed analyses of assessment strategies have been made on red deer and on toads.

The role of roaring in the assessment of fighting ability in red deer (*Cervus elaphus*) has been studied by Clutton-Brock & Albon (1979; see also Clutton-Brock *et al.* 1979). At the end of September or early October, hinds congregate in particular areas, where they are joined by stags which have spent the rest of the year in bachelor groups. Stags compete with one another for the possession of groups of hinds, or 'harems'. Individual stags between the ages of 7 and 11 years are

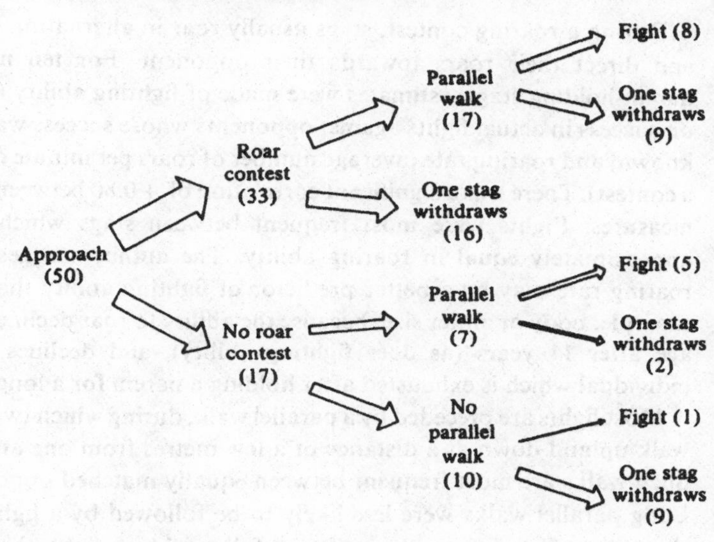


Figure 21. The course of contests between male red deer. (After Clutton-Brock & Albon, 1979.)

most successful in holding a harem. In this age range, a stag can hold a harem for 2–4 weeks; during this period he may have to fight another stag on average once in 5 days. There are large differences, within and between age classes, in the success of stags in holding harems. Stags lose up to 20% of their body weight during the rut. Fighting is potentially dangerous: 6% of stags were injured per year, indicating a chance of about 25% of serious injury per lifetime. Fighting is also costly for a harem holder because, during a protracted fight, his harem will be dispersed by younger stags.

Thus on the one hand fighting ability contributes to reproductive success, but on the other, fighting is costly and potentially dangerous; it is therefore to be expected that assessment will be used to settle contests. Figure 21 shows the course of 50 contests between two stags, of 6 years of age or more, which approached to within 100 m of each other. Of these, only 14 led to escalated fights, and in all but one of these cases escalation was preceded either by a roaring contest, or a parallel walk, or both. There are good grounds for thinking that both of these actions are concerned with the assessment of fighting ability.

During a roaring contest, stags usually roar in alternating bouts, and direct their roars towards their opponent. For ten mature harem-holding stags, estimates were made of fighting ability (based on success in actual fights against opponents whose success was also known) and roaring rate (average number of roars per minute during a contest). There was a significant correlation of +0.80 between these measures. Fights were most frequent between stags which were approximately equal in roaring ability. The authors suggest that roaring rate may be a better predictor of fighting ability than, for example, body or antler size, because the ability to roar declines with age after 11 years (as does fighting ability), and declines in an individual which is exhausted after holding a harem for a long time.

Most fights are preceded by a parallel walk, during which two stags walk up and down at a distance of a few metres from one another. Such walks are most frequent between equally matched opponents. Long parallel walks were less likely to be followed by a fight than short ones, but if a parallel walk was followed by a fight, then long fights tended to follow long walks. This suggests that if there is a substantial difference in fighting ability it will be detected by a long parallel walk, but that if a long walk fails to reveal such a difference the ensuing fight will be a long one, because the contestants are equally matched.

In the toad, *Bufo bufo*, Davies & Halliday (1978) have shown that depth of croak is used in assessment in fights between males. Females of this species come to ponds to spawn during one or two weeks in spring. The male clasps the female's back (amplexus), and may be carried around by her for several days before she eventually lays her eggs, which he fertilises externally. Females are usually mounted before they reach the pond. There is a considerable excess of males, partly because males spend the whole spawning period at the pond, whereas a female spends only a few days, and partly because there is, in any case, an excess of males of breeding age. Consequently, there is a great deal of wrestling between males for the possession of females.

In the laboratory, a larger male succeeded in displacing a smaller paired male in 10 out of 23 cases, whereas smaller intruders never succeeded in displacing larger paired males (0 out of 18 cases). Thus larger size is an advantage, but a smaller male in amplexus may succeed in resisting a larger intruder. Field observations confirmed

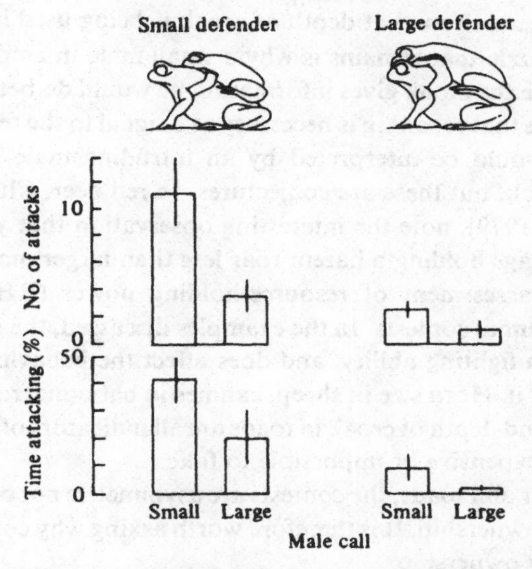


Figure 22. The effects of croak depth on attacks in the toad, *Bufo bufo*. Each attacker was used in two experiments. In one, he heard tape-recorded croaks of a small male and, in the other, croaks of a large male. The actual paired males were silenced by a rubber band passing through their mouths. One set of 12 males attacked small defenders and another 12 attacked large defenders. (After Davies & Halliday, 1978.)

this, and also showed that persistent fights are usually between a larger intruder and a smaller owner. When attacked, a paired male always croaks. The pitch of the croak is closely related to body size. The experiments illustrated in Figure 22 show that the pitch of croaks is used in assessment by attacking males. In these tests, medium-sized males were put into a tank containing a paired male and female. The paired male was either large or small, and had been silenced by a rubber band placed through the mouth. Each time the unpaired male touched the pair, croaks were played from a loudspeaker over the tank. These croaks were recorded either from a large or small male. The figure shows that the unpaired males were more likely to attack if the recorded call was that of a small rather than a large male; they were also more likely to attack if the actual owner was small, so croaking is not the only clue to size.

These experiments show that depth of croak is being used in size assessment. A puzzle that remains is why a small male in amplexus croaks at all, since the croak gives information he would do better to conceal. It may be that croaking is necessary as a signal to the female, or that silence would be interpreted by an intruding male as an invitation to attack, but these are conjectures. In red deer, Clutton-Brock & Albon (1979), note the interesting observation that young (5–6 years old) stags holding a harem roar less than larger males.

To conclude, assessment of resource-holding power (RHP) is taking place in animal contests. In the examples discussed, the signal is correlated with fighting ability, and does affect the behaviour of animals receiving it. Horn size in sheep, canines in baboons, roaring rate in red deer, and depth of croak in toads are all indicators of RHP which would be expensive or impossible to fake.

In both red deer and toads, the contests are asymmetric not only in RHP but also in ownership. It is therefore worth asking why contests are not settled by ownership.

From the payoff matrix of the Hawk–Dove–Bourgeois game (Table 12, p. 96) it follows that $B$ is only an ESS if $V < C$. In contrast, if it is worth risking injury ($-C$) to gain the resource ($V$), $H$ is the only ESS. In the red deer the risk of injury is relatively slight (6% per year) and in the toad it is non-existent. In both cases, a male which did not challenge owners might never achieve a mating. Hence, for these species it seems likely that $V > C$, so that Bourgeois is not a possible ESS.

I now turn to cases in which either Bourgeois or Assessor are possible ESS's. This is likely to be the case if the resource competed for is a territory, a burrow, a refuge or a web, because in such cases the alternative to obtaining a resource unit by fighting is not to do without, but to find or construct a resource unit for oneself.

An analysis of games in which more than one asymmetry exists has been undertaken by Hammerstein (1981). If two asymmetries exist, it might seem that the asymmetry with the larger payoff difference would be used in conventional settlement. This, however, turns out not to be the case. Hammerstein shows that it is possible for an aspect which does not affect payoffs to settle contests, even if a second aspect which does affect payoffs also exists. In other cases, one aspect may have such a strong effect on payoffs that it is necessarily used to decide contests.

One of the most extensive studies of contests in the field concerns male fiddler crabs (*Uca pugilator*) over burrows (Hyatt & Salmon, 1978). In 403 contests observed, the owner won in 349, and the intruder in 54, cases. However, in the latter cases, the intruder was larger in 50 contests and smaller only in one. Clearly, differences both in size and ownership are relevant to the outcome. Typically, ownership is taken as the arbiter, but a sufficiently large size difference can override this. It is difficult to analyse this further because little is known of the payoffs involved. Males have claws powerful enough to crush an opponent. Hyatt & Salmon, however, observed no injuries resulting from fights, so the main costs are in time and energy. In contrast, Jones (1980) reports that in a related species, *U. burgersi*, 25% of males had damage to their major chelae of the kind to be expected if it had occurred during fights. It is also hard to measure the value of the burrows (which are mating stations) over which the fights take place. Nevertheless, the basic conclusion, that both size differences and ownership influence outcomes, is well established and will probably prove to be typical.

The best illustration of the way in which asymmetries of size and ownership, and variations in payoffs, can influence contest behaviour is Riechert's (1978, 1979, 1981) study of the funnel-web spider *Agelenopsis aperta*. Females fight over webs, and associated territories. At any one time, a proportion of females lack webs. In a desert grassland study area, this proportion varied from 5 to 35%, depending on whether the previous season had been a favourable one for breeding. About 10% of webs changed ownership each day as a result of contests. Web owners gained on average 3.3 mg in weight per day, and non-owners lost 8.6 mg per day, mainly from water loss. At the end of the season, in July, most adults drown in the rains, but egg sacs are waterproof and hence eggs survive. Larger females lay more eggs; in severe seasons less successful females may lay no eggs.

Riechert studied contests in the field by releasing females close to web owners; 33 naturally occurring contests resembled these induced contests closely. A contest can pass through four phases of increasing costliness and risk as indicated below:

(i) 'Locating'; orienting movements, and palpation of the web, probably providing some information about relative size.

(ii) 'Signalling'; lengthy exchanges of vibratory and visual displays. Riechert (1978) found that during this phase, ultimate winners were less stereotyped in behaviour than ultimate losers.

(iii) 'Threat'; running or lunging towards an opponent.

(iv) 'Contact'; at its most extreme, this leads to two contestants rolling over the web, locked together. Death can result, but is rare (approximately 1% of contests); injury is not uncommon.

The median duration of disputes in the desert grassland site was 26 minutes, and some lasted many times as long; during a dispute, however, some 98% of the time is spent with both contestants motionless. The value of a web depends on local temperature and on the prevalence of prey. As Table 20 shows, the more valuable the web under dispute, the greater the duration of the contest.

There is evidence that only the owner knows the value of the web. Thus if the owner was removed and two intruders allowed to fight over a web, the correlation between web value and contest cost disappeared. The relevance of this to modelling the situation will appear below. Finally, contests in the desert grassland population lasted on average more than twice as long as in a riparian population. This is associated with the fact that in the grassland there are relatively few good sites, and these are all occupied, whereas in the riparian habitat many good sites are available and unoccupied. Experiments now in progress indicate that these differences are genetic (S.E. Riechert, personal communication).

Riechert distinguishes three general categories of contest:

(i) The shortest contests occur when the owner is substantially larger than the intruder. Indeed, 91% of contests were won by the larger spider. If the weight difference was greater than 30%, it was common for the larger spider to pass directly from 'locating' to 'threat' without a protracted signalling period, and for the smaller, particularly if it was the intruder, to retreat at once. Contests between approximately equal-sized pairs, in which the weight of the intruder had been doubled by glueing a lead weight to the abdomen, were usually won by the intruder, showing that web palpation and vibration does convey information about relative mass.

(ii) When there was only a small size difference ($< 10\%$), the contest lasted longer and involved a protracted signalling phase. If the size difference was less than 10%, the owner won in 90% of cases.

(iii) The longest contests, and the ones most likely to involve threat

Table 20. *Value of web and contest costs in* Agelenopsis aperta

| Site quality | Value of web (predicted reproduction in mg of eggs) | Mean contest cost (Joules × $10^6$ expended) |
|---|---|---|
| Poor | 70.4 | 6 |
| Average | 417 | 43 |
| Excellent | 653 | 140 |

and contact, occurred over webs of high value, when the owner was slightly smaller than the intruder.

Many features of these contests agree with the predictions of the simple models discussed earlier in this book. Both ownership and size assessment cues are being used, and, as expected in a war of attrition, the costs incurred increase with the value of the web. There are, however, many details remaining to be explained. For example, the lower stereotyping in the behaviour of ultimate winners is puzzling.

One particular question is worth further theoretical study. Why should the longest contests occur when the web is particularly valuable, and the owner is slightly smaller? We can get some insight from the following simple model:

(i) A fraction $p$ of all webs are of value $V$, and $1-p$ are of value $v$, where $V \geqslant v$.

(ii) Only two tactics are possible: Hawk, $H$ and Dove, $D$.

(iii) The owner and intruder differ in size, so that in an escalated contest there is a probability $x$ that the owner will win.

(iv) In an escalated contest, the winner receives $V$ or $v$ and the loser $-C$.

(v) The value of $x$ is known to both contestants, but the value of the web is known only to the owner.

(vi) Two Doves have equal chances of obtaining the web.

The main way in which this model fails to represent the situation in *Agelenopsis* is in assumption (ii). In practice, spiders are capable of a more graded set of behaviours. By 'knows' in assumption (v), I mean that the relevant parameter (for example, $x$) can influence behaviour. Presumably $x$ is known only after a period of assessment. Thus the

Table 21. *Payoff matrix for a model of contests in* Agelenopsis

| | | Intruder | |
|---|---|---|---|
| | | $H$ | $D$ |
| Owner | $H$ | $E(1-x)-Cx$<br>$Ex-C(1-x)$ | $0$<br>$E$ |
| | $CH$ | $p[V(1-x)-Cx]$ $+(1-p)v$<br>$p[Vx-C(1-x)]$ | $(1-p)v/2$<br>$(E+pV)/2$ |
| | $D$ | $E$<br>$0$ | $E/2$<br>$E/2$ |

statement that only two tactics, Hawk and Dove, are possible should be taken as a description of behaviour after an initial assessment phase, which is performed by all contestants.

I consider three possible strategies for the owner of a web:

(i) $H$; always escalate.

(ii) $D$; never escalate.

(iii) $CH$; 'Conditional Hawk'; adopt $H$ if the web is of value $V$, and $D$ if it is of value $v$.

We need consider only the strategies $H$ and $D$ for the intruder, because the intruder's tactics cannot depend on the value of the web, which is unknown. Note also that these strategies are conditional on a knowledge of $x$. Thus what we are seeking is a specification of what, knowing $x$, an owner will choose from $H$, $D$ and $CH$, and what, also knowing $x$, an intruder will choose from $H$ and $D$.

The payoff matrix is given in Table 21. In the table

$$pV+(1-p)v = E, \tag{9.1}$$

where $E$ is the expected value of a web to the intruder.

From the payoff matrix, one can write down the conditions for a particular tactic to be the best possible for the owner, given that the intruder chooses a particular tactic, and vice versa. For example, if the intruder chooses $H$, the best strategy for the owner is:

$$\left.\begin{array}{ll} & H \text{ if } Ex-C(1-x)>p[Vx-C(1-x)], \text{ or } x>C/(v+C), \\ & CH \text{ if } x<C/(v+C), \text{ and if } Vx-C(1-x)>0, \\ \text{or} & x>C/(V+C), \\ \text{and} & D \text{ if } x<C/V+C). \end{array}\right\} \quad (9.2)$$

If the intruder chooses *D*, it always pays the owner to choose *H*. Similar inequalities can be written down for the intruder's best choices, conditional on the owner's choice. If a pair of choices, *X* for the owner and *Y* for the intruder, is each the best reply to the other, then the strategy pair $(X, Y)$ is an ESS. This is most easily shown by a numerical example. Thus consider the case:

$$V = 2C;\ v = \tfrac{1}{2}C;\ p = \tfrac{1}{4}. \quad (9.3)$$

The best choices for owner and intruder, from inequalities such as (9.2), are shown in Figure 23. It turns out that there are two possible ESS's, type *A* and type *B*. Both have the following features in common:

(i) If the size difference is not great ($x \simeq \frac{1}{2}$), the owner's choice is conditional on the value of the web.

(ii) If the size difference is great, the larger wins without escalation.

(iii) There is a region, the 'escalation region', with $x \simeq \frac{1}{2}$, in which escalated contests occur when the web is valuable (value $V$), and the intruder wins without escalation when it is not (value $v$).

In all these respects, the model reflects the actual situation, with the proviso that an 'escalated' contest in the model corresponds to a prolonged contest, with much signalling and a higher risk of injury, in the real world. Riechert also observed that prolonged contests occurred most frequently when the owner was slightly smaller than the intruder. This too is a feature of the type *B* ESS, and remains true if the parameters are changed. It is not, however, a feature of type *A*. The exact symmetry of the escalation region either side of $x = \frac{1}{2}$ is peculiar to the particular parameter values chosen (i.e. $V/C = C/v$). As the parameter values are changed, the type *A* ESS has an escalation region which can lie either mainly above or mainly below $x = \frac{1}{2}$. I see no reason why one or other type of ESS should evolve. Hence the model can lead to the bias which Riechert observed, but does not predict it.

We are some way from a full analysis of *Agelenopsis* contests. Riechert has made progress in measuring the costs and benefits. The

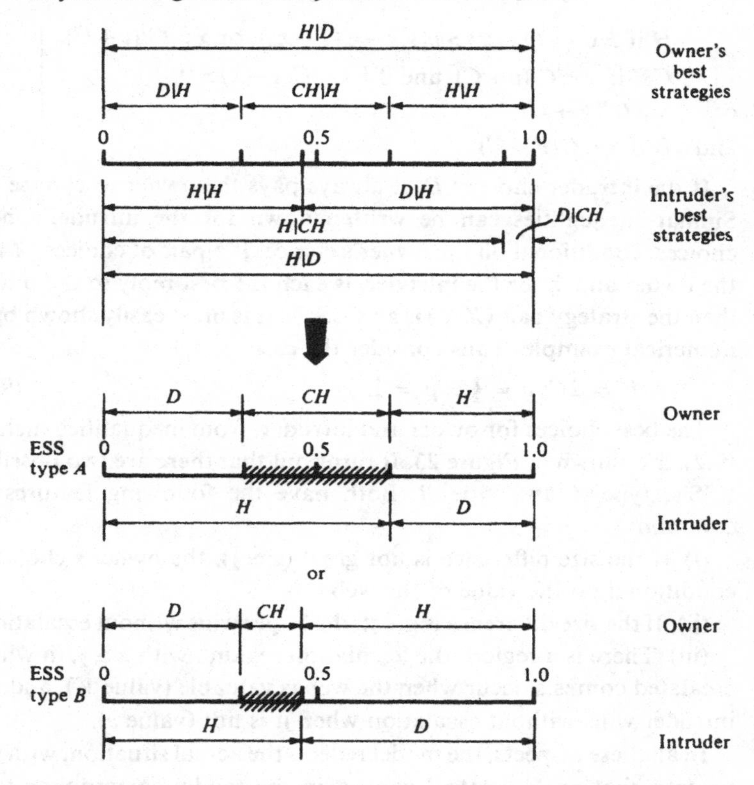


Figure 23. Model of contests in the spider *Agelenopsis*. The bold lines are scales of $x$, the probabilities that the owner will win an escalated contest. The upper diagram shows values of $x$ for which particular choices by one contestant are the best replies to choices by the other; for example, $H|D$ above the $x$ scale implies that $H$ is the best reply for the owner if the intruder chooses $D$, and so on. The two lower diagrams show the two possible ESS's derived from the upper diagram; the shaded area is the 'escalation range' (see text). The ranges shown are for $V=2C$, $v=\frac{1}{2}C$, $p=\frac{1}{4}$. For symbols, see text.

major difficulty which remains is that spiders are capable not just of two plays, Hawk and Dove, but of a range of behaviours. It is not easy to develop a model which takes this into account. A model in which the range of possible plays was fully continuous (as, for example, in the war of attrition) would not, I believe, give the right qualitative predictions. An appropriate model, perhaps, would be one which permitted a number of discrete levels of escalation.

One other example of a complex asymmetric contest will be described. On p. 32 it was explained how, after copulation, male dung flies remain on the backs of females, and how struggles may take place between a paired male, or 'owner', and an attacker. Sigurjónsdóttir & Parker (1981) observed 200 such struggles in the field, and recorded the duration of each contest, the outcome, the sizes of the two males and of the female, and the number of eggs remaining to be laid by the female, this last being a measure of the 'value' of the female.

In all but five cases, the attacker was larger than the owner. Since most approaches are followed by the retreat of the approaching male, without a struggle but after a display by the owner, it seems that size differences can be perceived, and that contests in which the attacker is smaller are decided conventionally in favour of the owner. Despite the larger size of the attacker when struggles do occur, the owner won on almost 75% of occasions. Thus the owner has a positional advantage, a complication which does not arise in *Agelenopsis* contests.

In analysing the data, a serious difficulty is that, for any particular contest, one can measure only the time for which the loser was prepared to continue; of the winner, one knows only that it was prepared to continue for longer than that. Nevertheless, the following conclusions can be drawn:

(i) A male's persistence is influenced, not by its absolute size, but by its size relative to its opponent. As the relative size of the owner increases, the persistence of the attacker decreases and of the owner probably increases.

(ii) The larger the female, the greater the probability of takeover, probably because the attacker is more persistent if the female is large. Since large females lay more eggs, size is a predictor of the value of a female to the attacker.

(iii) The owner's persistence was negatively correlated with the length of time it had been guarding, and therefore positively correlated with the number of eggs remaining to be laid. The attacker's persistence was uncorrelated with either of these variables; an attacker has no way of estimating the value of a female other than by her size.

This example resembles the *Agelenopsis* one in many ways. The strategy set is continuous rather than discrete. There are asymmetries both of size and of ownership. Owners have information about the value of the resource not available to intruders. I suspect that these features may turn out to be typical of pairwise contests between animals over indivisible resources.

# 10 *Asymmetric games – III. Sex and generation games*

## A Some theoretical considerations

We owe primarily to Trivers (1972, 1974) the recognition that conflicts of interest may arise between members of a species which differ in age or sex. The meaning of the word 'conflict' in this context can best be illustrated by the example of parent–offspring conflict over weaning time (Trivers, 1974). Suppose that the age at weaning would be $X_M$ if it were determined by genes expressed only in the mother, and would be $X_C$ if it were determined by genes expressed only in the child, then we can speak of conflict if $X_M \neq X_C$. Conflict can similarly occur between males and females, for example over whether mating shall take place or over the extent of parental care provided by each. In a similar spirit, one can speak of conflict between genes expressed in zygotes and gametes, or between chromosomal and cytoplasmic genes (Cosmides & Tooby, 1981; Eberhard, 1980*b*).

In this chapter I discuss how game theory models can be applied to such problems. The methods are more useful for analysing conflicts between males and females than between parents and offspring. The latter type of conflict requires an explicit genetic model because, by definition, offspring are related to their parents, and models of games between sexual relatives are not easy to develop. For the same reason, genetic models are needed for conflicts over mating between relatives (Packer, 1977*a*; Parker, 1979) or over the sex ratio between queens and workers (Trivers & Hare, 1976). However, conflicts over mating and parental care in random-mating populations raise no such difficulties, and can usefully be tackled by game theory.

Logically, there is no distinction between an asymmetric game in which players in roles 1 and 2 each have the same two options, say *A* and *B*, open to them, but receive different payoffs according to their roles, and a game in which the player in role 1 has options *A* and *B*

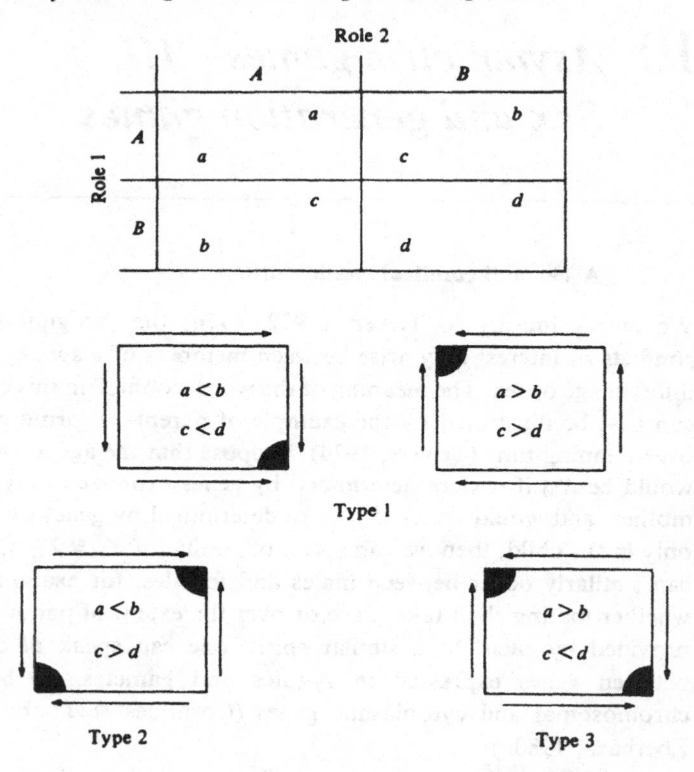


**Figure 24. Types of ESS for a game with an uncorrelated asymmetry. Shaded areas indicate stable points.**

and in role 2 has options *R* and *S*. The important logical distinction is between games with an uncorrelated asymmetry (p. 95) on the one hand, and games with differences either of payoffs or of choices available on the other. Thus the distinction between the games discussed in this chapter and the last is biological rather than mathematical.

The possible outcomes for a game with an uncorrelated asymmetry are shown in Figure 24. Note that if the asymmetric strategy 'In role 1, *A*; in role 2, *B*' is an ESS, so is the opposite strategy 'In role 1, *B*; in role 2, *A*'.

Consider now the game in which either the payoffs alone, or the

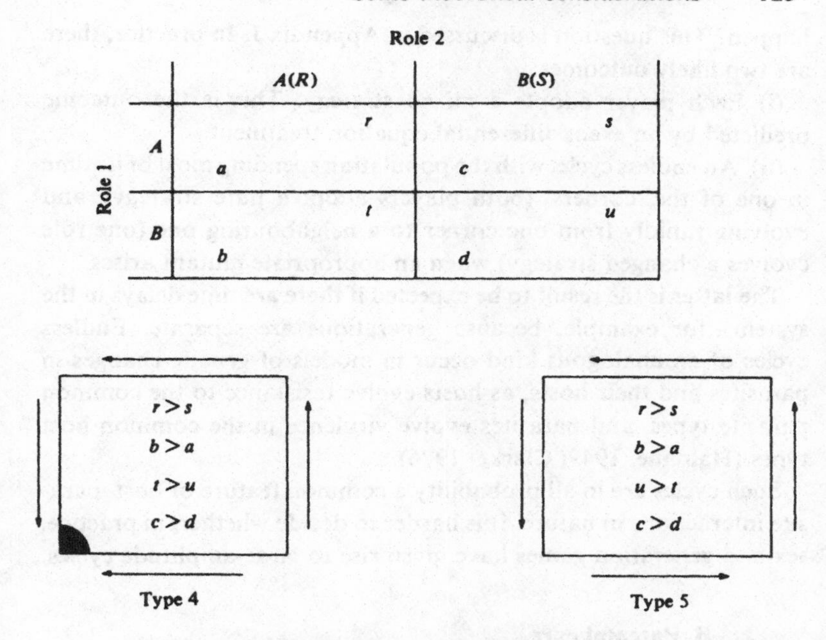


Figure 25. Additional types of ESS when the payoffs are different in the two roles.

choices and payoffs, are different in the two roles. Two additional, qualitatively new types of solution now become possible (Figure 25).

Games such as those discussed in the last chapter, in which the same tactics are available in both roles but payoffs are different, can easily give rise to situations of type 4, Figure 25. Thus we found that in some cases both a common-sense and a paradoxical ESS existed (type 2, Figure 24), but that in others only the common-sense strategy was stable (type 4, Figure 25). Even when two ESS's exist, it is usually clear that one is more likely to arise than the other: i.e. one is 'common-sense' and the other 'paradoxical'. In contrast, when the tactics available are different in the two roles, it is often difficult to decide which of the two ESS's is the more likely to arise. This point is illustrated below by the example of parental care.

Cyclical solutions (type 5, Figure 25) are also most likely to arise when the choices available in the two roles are different. Suppose, however, that the payoffs obey the appropriate inequalities, what will happen? This question is discussed in Appendix J. In practice, there are two likely outcomes:

(i) Each player adopts a mixed strategy. This is the outcome predicted by an exact differential equation treatment.

(ii) An endless cycle, with the population spending most of its time in one of the 'corners' (both players adopt a pure strategy), and evolving rapidly from one corner to a neighbouring one (one role evolves a changed strategy) when an appropriate mutant arises.

The latter is the result to be expected if there are time delays in the system – for example, because generations are separate. Endless cycles of an analogous kind occur in models of genetic changes in parasites and their hosts, as hosts evolve resistance to the common parasite types, and parasites evolve virulence in the common host types (Haldane, 1949; Clarke, 1976).

Such cycles are in all probability a common feature of host–parasite interactions in nature. It is harder to decide whether, in practice, sex and generation games have given rise to large-amplitude cycles.

## B Parental care

A simple game theory model of parental care (modified from Maynard Smith, 1977) takes into account the following effects:

(i) The value of parental care by one or two parents.

(ii) The chance that a male mates again, depending on whether or not he guards the offspring from his first mating.

(iii) The effect of parental care on the number of eggs the female can lay.

Thus assume that for both males and females there is a choice either of 'Guarding', *G*, or of 'Deserting', *D*. (Models in which the duration of parental care is variable were also considered by Maynard Smith (1977) and by Grafen & Sibly (1978).)

Let $P_0$, $P_1$, $P_2$ be the probabilities that an egg will survive if it is cared for by 0, 1 or 2 parents respectively. $P_0 \leq P_1 \leq P_2$.

Let $p$, $p'$ be the probabilities that a male mates a second female if, respectively, he deserts or he guards. (In the earlier treatment I assumed $p' = 0$, which is often not the case.)

Let $V$, $v$ be the numbers of eggs laid by a female if, respectively, she deserts or she guards; $V \geq v$.

Table 22. *Payoffs in the parental care game*

| | | Female | |
|---|---|---|---|
| | | *G* | *D* |
| Male | *G* | $vP_2$ / $vP_2(1+p')$ | $VP_1$ / $VP_1(1+p')$ |
| | *D* | $vP_1$ / $vP_1(1+p)$ | $VP_0$ / $VP_0(1+p)$ |

The probability that a male is the father of the offspring of a female with which he mates is assumed to be independent of whether he guards or deserts. I will return to this assumption later.

With these assumptions, the expected fitnesses are given in Table 22. There are four possible ESS's, as follows:

*ESS 1.* Both *G*.
$vP_2 > VP_1$, or female deserts,
and $P_2(1+p') > P_1(1+p)$, or male deserts.

*ESS 2.* Male *G*, female *D*.
$VP_1 > vP_2$, or female guards,
and $P_1(1+p') > P_0(1+p)$, or male deserts.

*ESS 3.* Male *D*, female *G*.
$vP_1 > VP_0$, or female deserts,
and $P_1(1+p) > P_2(1+p')$, or male guards.

*ESS 4.* Both *D*.
$VP_0 > vP_1$, or female guards,
and $P_0(1+p) > P_1(1+p')$, or male guards.

If $P_2 \gg P_1$ (two parents much better than one), and $p$ not much greater than $p'$ (desertion does not confer great advantages in re-mating), then both parents will guard.

If $P_0$ is not greatly less than $P_1$, then it is likely that both parents will desert.

Single-parent guarding is favoured if $P_2 \simeq P_1 \gg P_0$. Male desertion is favoured if $p > p'$ (desertion favours a second mating). Female desertion is favoured if $V > v$ (guarding uses up resources which could otherwise be allocated to eggs). However, it is easy to find parameter values for which both ESS's 2 and 3 exist; only one parent guards, but it can be either the male or female. As suggested above, situations with two alternative asymmetric ESS's (type 2, Figure 24) are likely to arise in games of this type. Which ESS will be reached depends on initial conditions, and it may be difficult in particular cases to reconstruct the ancestral behaviour from which the present behaviour evolved.

Ridley (1978) has reviewed cases in which only the male cares. The strongest association is with external fertilisation. Of 55 families with male care, 35 have external fertilisation, and 20 either have internal fertilisation or the eggs are released after the sperm. This contrasts with the great excess of cases of internal fertilisation associated with female care. Two reasons have been suggested. Dawkins & Carlisle (1976) argued that, if one parent is adequate to care for the young, then the sex which deserts will be the one which is first free to do so. Hence, with internal fertilisation the male will desert, whereas with external fertilisation and synchronous gamete production either sex might desert. This may well account for a substantial part of the association between external fertilisation and male care; if the male is not there when the eggs are laid, he cannot care for them.

The second reason which has been proposed is that with external fertilisation a male has greater confidence in paternity (Trivers, 1972). There is obvious force in this argument; a male is unlikely to guard offspring which are not his own. Why, then, was confidence of paternity omitted from the model above? In fact, it was not. When comparing species, confidence of paternity is inversely related to the chance of a second mating. Thus with a 1 : 1 sex ratio, if females on average mate ten times, each male can expect ten matings, and will father one-tenth of the offspring of each female with which he mates. The model predicts that species with a high chance of further matings will show male desertion. Equivalently, species with a low confidence of paternity will show male desertion.

There is more to be said, however, about mode of fertilisation and confidence of paternity. Within a species, selection will favour behaviour by males which increases the probability that they will father the offspring produced by females with which they mate; if males do guard offspring, selection will favour males which guard offspring from those matings for which they have the highest confidence of paternity. With internal fertilisation, the former objectives can be achieved by post-copulatory guarding (Parker, 1974*c*). If, after mating, a male guards the female until the eggs are laid, he is almost in the same position as an externally-fertilising male. Consequently, in species with internal fertilisation, paternal care is usually associated with post-copulatory guarding. An alternative is that paternal care may evolve in monogamous species from biparental care.

Two other links between the model and comparative data are worth mentioning. First, it is often the case that a male which cares for his offspring also defends a territory, and sometimes a nest, in which females can lay eggs (Trivers, 1972; Ridley, 1978). In such cases, a male which guards his offspring may at the same time increase his chance of further matings ($p' > p$); clearly, this favours male care. Secondly, male care is associated with species in which the number of eggs a female can lay would be substantially reduced if she had to devote resources to parental care ($V \gg v$). This is not usually the case in birds, in which the number of eggs laid is limited by the number which can be incubated, or by the number of young which can be fed (Lack, 1968). Accordingly, male-only care is rare in birds. It is, however, found in some species (e.g. the rhea, Bruning, 1973; the mallee-fowl, Frith, 1962; the roadrunner, S.L. Vehrencamp, personal communication), in which there are good reasons to think that egg production by females is resource-limited.

The preceding discussion has been concerned with male–female conflicts over parental care. Parker (1979) has discussed other contexts in which conflict can arise; although his discussion is confined to insects, it has a general relevance. Because of the discrepancy in investment in gametes between males and females, males are likely to be less discriminating over mating than females (Bateman, 1948; Maynard Smith, 1956; Trivers, 1972). One particular form that this conflict can take has been pointed out both by Parker (1979) and Packer (1977*a*). The level of inbreeding depression beyond which it would no longer pay a male to mate with his sister is more extreme than the level at which it would not pay a female to mate with her brother. Consequently there is a range of levels for which selection would favour incestuous mating by one sex but not the other. The same will be true for distant outcrossing.

Fighting between males over females can be damaging to females. Parker reports that during fights between male dung flies (*Scatophaga stercoraria*), females can be injured, or even drowned in the dung. Female toads can also be drowned during inter-male fights. If, as in lions (Bertram, 1976) and langurs (Hrdy, 1974), male takeover results in infanticide, females may oppose takeover. Pre- and post-copulatory guarding is common in arthropods (Parker, 1974*c*). The advantage to the male, in achieving matings and in ensuring paternity, is obvious, but the practice must impose some cost on females, particularly if the female must carry the male around.

### C Games with cyclical dynamics

Dawkins (1976) considers the following imaginary game. Suppose that the successful raising of an offspring is worth +15 to each parent. The cost of raising an offspring is −20, which can be borne by one parent only, or shared equally between two. The cost of a long courtship is −3 to both participants. Females can be 'coy' or 'fast'; males can be 'faithful' or 'philanderer'. Coy females insist on a long courtship, whereas fast females do not; all females care for the offspring they produce. Faithful males are willing, if necessary, to engage in a long courtship, and also care for the offspring. Philanderers are not prepared to engage in a long courtship, and do not care for their offspring. With this assumption, the payoff matrix is shown in Table 23.

The characteristic feature of this matrix is its cyclical character. That is:

If females are coy, it pays males to be faithful.

If males are faithful, it pays females to be fast.

If females are fast, it pays males to philander.

If males philander, it pays females to be coy.

Thus we have come full circle. The dynamics of such games are discussed in Appendix J. Oscillations are certain, but whether they are divergent or convergent will depend on details of the genetics. If divergent, a population would in practice spend much of its time fixed

Table 23. *Dawkins' (1976) battle of the sexes*

| | | Female | |
|---|---|---|---|
| | | Coy | Fast |
| Male | Faithful | 2 / 2 | 5 / 5 |
| | Philanderer | 0 / 0 | 15 / −5 |

for a particular pair of strategies. Such a population would be vulnerable to invasion by mutants. Hence, when appropriate mutations occurred it would evolve rather rapidly to a new pair of strategies, and would then again have to wait for new mutations.

Dawkins' game was an imaginary one. Parker (1979), using explicit genetic models, has suggested that similar cycles could arise from parent–offspring conflict. I am unable to offer illustrative examples, or evidence that such cycles occur. The difficulty is that such cycles, if they occurred, would be of too long a period to be readily observed. Simple models, however, lead so directly to the conclusion that oscillations will occur that it would be wise to be on the lookout for them.

### D Sexual selection

The most discussed example of an asymmetric game in which different strategy sets are available to the two players is that of sexual selection. Together with the problem of the sex ratio, it is also the first biological phenomenon to have been discussed from the standpoint, although not of course in the terminology, of evolutionary game theory, in Fisher's *The Genetical Theory of Natural Selection* (1930).

Fisher's argument can be put as follows. Suppose that at some time, most of the females in a population prefer as mates males with an extreme value of some phenotypic trait; for example, they prefer the male with the longest tail. Then, if the species is polygynous, or if early mating confers an advantage, males with longer tails will be fitter. Further, if tail length is heritable, females which choose males with longer tails will have sons with longer tails, and therefore will have more grandchildren. Hence, there will be runaway selection for longer tails in males, and choice of longer tails in females, which will continue until natural selection against excessive tail length arrests it.

The final state of such a population, in which male tail length is so great that it significantly reduces the chance of survival, can be seen as an asymmetric ESS. It would not pay a male to have a shorter tail, because he would fail to get a mate, and it would not pay a female to choose a male with a short tail because, if she did, her sons would fail to get mates.

How could such a process be initiated? Fisher recognised that at first there would have to be a correlation between the selected trait and high fitness; males with longer tails would have to be fitter. An alternative starting point could be as follows. Suppose that two previously isolated species, *A* and *B*, come to occupy the same area, and that some hybridisation takes place. Then, as Dobzhansky (1951) argued, selection will favour reproductive isolation. If, on average, males of species *A* had tails longer than those of species *B*, selection would favour females of species *A* which preferred males with longer tails. Fisher's argument suggests that, once started, the process could continue long after any need for species isolation was past.

Ironically, although Fisher's argument was essentially game theoretic, it turns out to be difficult if not impossible to give a more formal game-theoretic analysis; instead, the problem seems to be one which demands a formal population-genetics treatment. Such a treatment has been given in an important but difficult paper by Lande (1981). The basic assumptions and conclusions will now be described. When this has been done, it will be easier to see why a game-theoretic analysis runs into difficulties.

Let $z$ be a measure of some male trait, such as tail length. The trait is expressed only in males. The probability that a male survives to breed is $\phi(z)$; it is supposed that there is some optimal value, $z_{OPT}$, with survival probability falling off to either side.

It is supposed that each female has some 'preference', measured by $y$. The function $\psi(z|y)$ is proportional to the chance that a male of phenotype $z$ will be chosen by a female of preference $y$. Thus if there are two kinds of males, $z_1$ and $z_2$, in proportions $p$ and $1-p$, the probability that a $y$ female will mate a $z_1$ male is

$$p\psi(z_1|y)/[p\psi(z_1|y)+(1-p)\,\psi(z_2|y)].$$

Lande considers three kinds of preference function:

(i) Directional preference; $\psi(z|y)=\exp(yz)$. Thus all females prefer males with large $z$, but they differ in the degree of discrimination.

(ii) Absolute preference; $\psi(z|y)=\exp[-k(z-y)^2]$, where $k=\text{const}$. Thus a $y$ female prefers a male of tail length $z=y$, and her willingness to mate falls off either side of this value.

(iii) Relative preference; $\psi(z|y)=\exp\{-k[z-(\bar{z}+y)]^2\}$. Thus a $y$ female prefers a male with a tail length greater than the population mean, $\bar{z}$, by an amount $y$.

The fitness of a male, $z$, then depends both on his chance of surviving, $\phi(z)$, and on his mating success, which depends in turn on the distribution of female preference, $y$, as well as on $z$.

It is assumed that all females can mate, and that there is no male parental investment. Hence, there is no direct selection on females, since all can mate and the genes influencing tail length are not expressed in females. Why, then, should genes affecting $y$ alter in frequency? Essentially, because there is a genetic covariance between $y$ and $z$ caused by assortative mating. In other words, males with genes for high $z$ will tend to have genes for high $y$, and vice versa.

The reasons for this genetic covariance are worth spelling out in more detail. Consider a female with high $y$. She will pass genes for high $y$ to her offspring; also, because she mates with a male with high $z$, those offspring will receive genes for high $z$ from their father. Thus part of the covariance arises from assortative mating in the immediately preceding generation. There is, however, a further effect, arising from assortative mating in earlier generations, which causes linkage disequilibrium, so that gametes with genes for high $y$ tend also to carry genes for high $z$.

In Lande's model, then, female preference, $y$, evolves only because selection on males changes $z$, and there is non-random association between genes affecting $y$ and $z$. (Genetic covariance could also arise

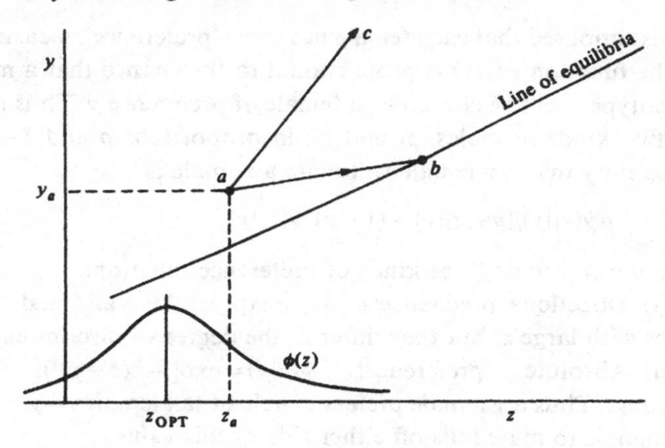


Figure 26. Conclusions of Lande's (1981) model of sexual selection. For explanation, see text.

from pleiotropy, but since the traits concerned are male morphology and female preference it is reasonable to ignore this possibility.) The first conclusion is that there exists, not a single equilibrium point, but a line of possible equilibria (Figure 26); Kirkpatrick (1982) has drawn a similar conclusion from a two-locus model. Thus for each value of mean female preference, $\bar{y}$, there is a corresponding mean male tail length $\bar{z}$. Mean tail length need not correspond to the optimum length for survival. If a population is on the line of equilibria, there is no net force tending to drive it along the line.

There is one feature of the equilibrium which is at first sight counter-intuitive. Consider the case in which females have an absolute preference, $\bar{y}$. The fittest male zygotes have the phenotype $\bar{z}$; the net effect of survival and mating selection is normalising. One might think, therefore, that the type of male most preferred by females would also be $\bar{z}$; that is, that $\bar{y} = \bar{z}$. This, however, is not the case. In fact, $\bar{y} > \bar{z}$. The reason is that, if a female is to maximise her chance of mating with a male of phenotype $\bar{z}$, she must prefer males with a higher value of $z$, to counteract the fact that most surviving males have values of $z$ lower than $\bar{z}$.

An intuitive feel for how this equilibrium is maintained is given in Figure 27, in which it is assumed that all females have the same absolute preference $\bar{y}$. The initial distribution of $z$ in zygotes, before

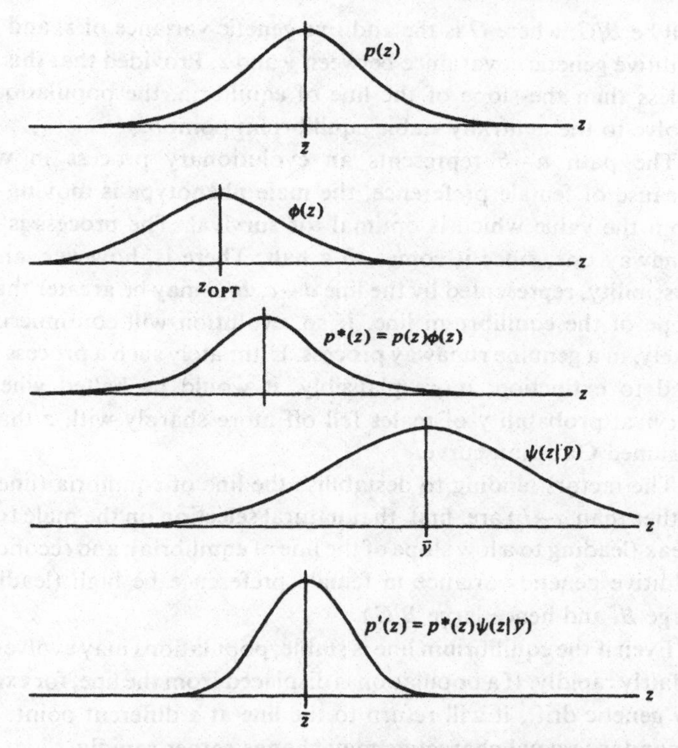


Figure 27. The maintenance of an equilibrium in Lande's (1981) model of sexual selection. For explanation, see text.

selection, is $p(z)$, with mean value $\bar{z}$. The distribution after survival selection is proportional to $p^*(z)=p(z)\phi(z)$. The function $\psi(z|\bar{y})$ measures the mating success of males, so $p'(z)=p^*(z)\psi(z|\bar{y})$ gives the proportional frequency distribution of mating males. At equilibrium the means of $p(z)$ and $p'(z)$ must be the same; the variance of $p'(z)$ is lower, the equilibrium variance in Lande's model being maintained by mutation.

How will the population evolve when not on the line? Suppose that the population lies above the line (point $a$, Figure 26). Then the degree of female preference, $y_a$, is greater than is needed to maintain the existing value of tail length, $z_a$. Hence $z$ will increase under sexual selection, and $y$ will increase because of the genetic covariance between them. Lande shows that the slope of the evolutionary path will be $B/G$, where $G$ is the additive genetic variance of $z$, and $B$ the additive genetic covariance between $y$ and $z$. Provided that this slope is less than the slope of the line of equilibria, the population will evolve to the neutrally stable equilibrium point $b$.

The path $a \to b$ represents an evolutionary process in which, because of female preference, the male phenotype is moving away from the value which is optimal for survival. The process is not a runaway one, since it comes to a halt. There is, however, another possibility, represented by the line $a \to c$. $B/G$ may be greater than the slope of the equilibrium line. If so, evolution will continue indefinitely, in a genuine runaway process. Ultimately such a process could lead to extinction; more plausibly, it would be halted when the survival probability of males fell off more sharply with $z$ than the assumed Gaussian curve.

The factors tending to destabilise the line of equilibria (line $a \to c$ rather than $a \to b$) are, first, that natural selection on the male trait be weak (leading to a low slope of the line of equilibria), and second, that additive genetic variance in female preference be high (leading to large $B$, and hence large $B/G$).

Even if the equilibrium line is stable, populations may evolve along it fairly rapidly. If a population is displaced from the line, for example by genetic drift, it will return to the line at a different point. Thus, secondary sexual characters may change rather rapidly.

Having summarised Lande's population genetics model, it is easier to see the difficulty of a game-theoretic analysis. Thus suppose payoffs were measured simply in numbers of offspring produced. All females would be equally fit, and hence $y$ would not change. The problem would reduce to the trivial one of asking what value of $z$ will evolve for a fixed value of $y$. To get round this, one might measure female payoffs in terms of numbers of grandchildren produced, since then one would be taking into account the fitness differences between their sons. This approach is more promising, but runs into two obvious snags.

(i) Suppose one analyses the stability of some population $y^*z^*$ by considering invasion by mutants $yz^*$ and $y^*z$. The attempt breaks down, because $yz^*$ females are identical in fitness to $y^*z^*$ females; they do not differ if only one kind of male exists. Thus one is forced to analyse invasion by $y$ and $z$ mutants simultaneously.

(ii) To calculate the fitness of a female's sons, one has in effect to know the genetic covariance between $y$ and $z$.

To overcome these difficulties would, in effect, be to write down a full population genetics model. I have spent time on the problem of sexual selection mainly to bring out the fact that, even if we are interested in the evolution of phenotypes and have no special knowledge of the nature of the genetic variance of those phenotypes, there is no guarantee that a game theory model will be adequate. We may be driven to the more difficult task of analysing an explicit genetic model.

## E Games with alternate moves

Some asymmetric contests are best understood by imagining that one contestant makes a first 'move', and that the other then selects a best reply to the move. For example, A. Grafen (personal communication) has analysed the evolution of the sex ratio in social hymenoptera in this way. Trivers & Hare (1976) asked what sex ratio would evolve if the queen was in control, and if the workers were in control; since, in ants, the actual sex ratios are closer to the 3:1 preferred by the workers, they concluded that the sex ratio is controlled by the workers. Grafen argues that the queen is able to determine the numbers of male and female eggs she produces, subject to a limit on the total numbers, and the workers to determine the numbers of male and female reproductives and workers they will raise, subject again to a limit on the total number and also on the numbers of male and female eggs the queen has laid. He seeks an ESS, such that neither queen nor workers can increase their inclusive fitness given the strategy adopted by the other. The analysis is complex and will not be given in detail here. The essential point is that, in order to find the queen's best strategy, it is necessary to decide, for each strategy she could adopt, what the workers would do in response, and to select that strategy which maximises her inclusive fitness, in the light of that response.

A simpler example of the same approach is Vehrencamp's (1979) analysis of reproductive skew in social groups. Suppose there is a group of $k$ potential reproductives, of which one is dominant and the others subordinates. If the $k$ individuals remain as a group, the total

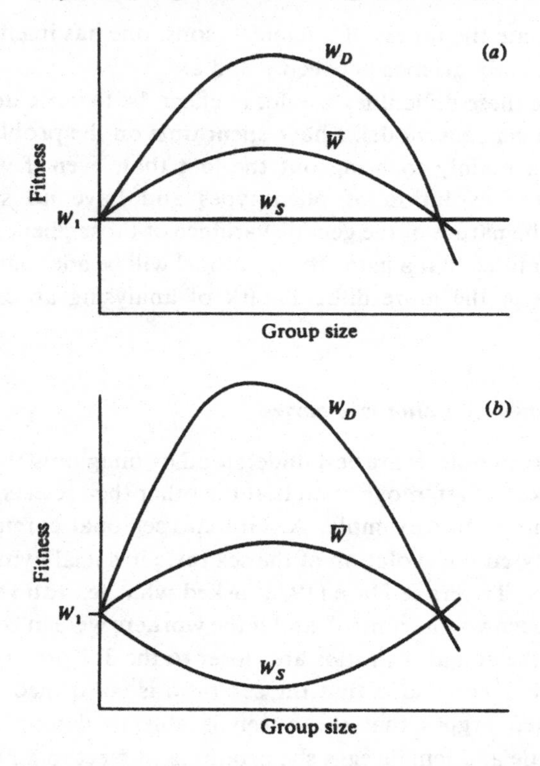


Figure 28. Skew in reproductive success in groups of (*a*) unrelated and (*b*) related individuals. $W_1$, fitness of solitary breeder; $W_S$, $W_D$, fitnesses of subordinate and dominant individuals, respectively; $\bar{W}$, mean fitness of group breeders. (After Vehrencamp, 1979.)

number of offspring produced is $k\bar{W}$ where $\bar{W}$ is the mean fitness of group breeders; if they split up and breed indvidually, each can produce $W_1$ offspring. The dominant can produce any number, $W_D$, of direct offspring, up to $k\bar{W}$, the remaining $k\bar{W} - W_D$ being divided equally between the $k-1$ subordinates. A subordinate can leave and breed alone. Vehrencamp assumes that if one subordinate leaves, they all do; it is also possible to work out stable strategies if subordinates leave one at a time.

In analysing the situation, we suppose that the dominant makes the first 'move' by selecting a value $W_D$. The subordinates then make their move, either staying or leaving. The dominant will select a value of $W_D$ which maximises its fitness, allowing for the behaviour of the subordinates. If the individuals are unrelated, the ESS is easy to see (Figure 28*a*). The dominant will not suppress the subordinates below $W_1$, or they will leave. If $\bar{W} < W_1$, individuals will be solitary. If $\bar{W} > W_1$, they will remain as a group, with each subordinate producing $W_1$ offspring (or slightly above $W_1$), and the dominant $k\bar{W} - (k-1)W_1$ offspring. If the dominant is able to influence the size of the group (e.g. if subordinates leave one at a time), it will choose that value of $k$ which maximises $k\bar{W} - (k-1)W_1$.

Suppose now that the individuals are related, with coefficient of relationship $r$. Let $W_0$ be the number of direct offspring produced by a subordinate in a group; it must be such that it would not pay the subordinate to leave and breed alone. To find $W_0$, we calculate $IF_S$ and $IF_G$, the 'inclusive fitness' of a subordinate breeding alone and in a group, respectively. What is calculated, in each case, is the total number of copies, identical by descent, of a gene $A$ present in a subordinate which are passed on to offspring by the members of the group as a whole. This is not strictly the inclusive fitness in Hamilton's (1964) sense. However, if $A$ and $A'$ are genes causing a subordinate to stay in the group or to leave, respectively, then if $IF_G > IF_S$ it follows that the frequency of $A$ will increase relative to $A'$. Thus:

$$IF_S = W_1 + r(k-1)W_1$$
$$IF_G = W_0 + r(k\bar{W} - W_0).$$

The minimum value of $W_0$, such that it is just not worth while to leave, occurs when $IF_S = IF_G$,

or $$W_0 = W_1 - [rk(\bar{W} - W_1)/(1-r)].$$

This is illustrated in Figure 28*b*. For groups to exist at all, it is necessary that $\bar{W} > W_1$; that is, there must be an overall advantage to group breeding. This is true whether or not group members are related. If there are groups, the skew in breeding success between dominants and subordinates will increase with $r$, and with $\bar{W} - W_1$. In unrelated groups, subordinates must do as well as they would do on their own ($W_0 \geqslant W_1$); in related groups, subordinates will do less well, and in the limit may not reproduce at all.

# 11 *Life history strategies and the size game*

There is a substantial body of work on the evolution of life history strategies; for reviews, see Stearns (1976) and Charlesworth (1980, ch. 5). The problem concerns the allocation of resources between survival, growth and reproduction. Most previous analyses have treated it as a 'game against nature'; that is, as a problem in optimisation with fixed constraints on rates of growth, mortality and fecundity. An exception is the analysis by Mirmirani & Oster (1978) of plant growth. The rate of growth of a plant will depend on its own size, and also on the size of its neighbours and the intensity of competition from them for light, water and nutrients. The optimal strategy – i.e. that strategy which allocates resources between growth and reproduction so as to maximise total reproduction output – will be different for a plant growing by itself and for one growing in competition with others. In the latter case, we have to find an evolutionarily stable growth strategy. Mirmirani & Oster consider the additional complication that neighbours may be relatives, but this possibility will be ignored in what follows.

The search for evolutionarily stable life history strategies will be difficult. In this chapter, I consider only one special case. The animal I had in mind when formulating this model was the male red deer. In the late aùtumn, the stags hold harems of up to 15 hinds. Fighting occurs between stags, and success in fighting enables a stag to hold a larger harem for longer. Stags less than six years old do not enter the rut, though they may perhaps achieve some matings by sneaking. Stags above that age participate in the rut each year. During the rut they use up most of their fat reserves, and in consequence grow little or not at all.

The model derived from this life history pattern is as follows. Individuals grow, but do not reproduce, up to age $x$, at which time they have reached size $m$. The larger $x$, the smaller the probability they will survive to that age. It is assumed that the probability of

Table 24. *The opponent-independent costs game*

| Player | $A$ | $B$ |
|---|---|---|
| Bid | $m_A$ | $m_B$ |
| Payoffs | $V-m_A$ | $-m_B$, if $m_A > m_B$ |
| | $V/2-m_A$ | $V/2-m_B$, if $m_A = m_B$ |
| | $-m_A$ | $V-m_B$, if $m_A < m_B$ |

survival, and the size reached, depend only on $x$, and not on what other members of the population are doing; in this respect, the model is simpler than the one considered by Mirmirani & Oster (1978). After age $x$ there is no further growth. Breeding success each year depends on $m$; more precisely, it depends on the size of the individual relative to other members of the population. What is the evolutionarily stable value of $m$?

A model similar to this was analysed by Parker (1979), in a different context. He considered contests between males and females, rather than between males for females. Thus the success of a male depended on his size relative to that of a female, and vice versa; i.e. he considered an asymmetric version of the present game. He concluded that no ESS existed, so that his system cycled continuously. It may be that this instability is an example of the cyclical instability discussed on p. 130 and Appendix J. An alternative possibility, as will emerge in a moment, is that the instability is a feature of the model also in its symmetric form.

Thus Parker (1979) sees his model as a modification of the war of attrition, in which the winner must pay his full bid, and not just the bid of the loser; if an animal grows to a size larger than was necessary to win a contest, it cannot demand some of its money back. Parker calls this the 'opponent-independent costs' game. The payoff matrix for the game in its symmetric form is shown in Table 24.

Haigh & Rose (1980) have shown that this game has no ESS; hence I suggested above that the instability that Parker found by simulation may also be a feature of the symmetric version of his model. Haigh & Rose suggest a modification of the model which does have an ESS. However, I prefer here to make more drastic changes. The game of Table 24 seems to me unsatisfactory for two reasons:

(i) The fitness of an individual should be the *product* of its probability of surviving to breed and its fecundity if it does breed. It cannot be adequately represented by subtracting a cost, $m_A$ or $m_B$, from a reward $V$.

(ii) Fecundity does not depend on success in pairwise contests.

The model I propose is as follows. The size distribution of the breeding population is $p(y)$, where

$$\int_0^\infty p(y)\,\mathrm{d}y = 1.$$

An individual of size $m$ will be larger than a random opponent with probability

$$z = \int_0^m p(y)\,\mathrm{d}y.$$

Let the probability of surviving to size $y$ be $s(y)$. If an individual of size $m$ participates in $k$ pairwise contests against random opponents, gaining $V$ for each victory, its overall fitness is

$$W(m) = kVs(y)\,z.$$

In many competitive situations (including leks, contests for harems, contests for positions in hierarchies), however, there is no reason why $W(m)$ should be a linear function of $z$. Thus a more general model is

$$W(m) = s(y)\,V(z), \tag{11.1}$$

where $V(z)$ is a non-linear function of $z$.

Given the forms of $s(y)$ and $V(z)$, we would like to find a distribution $p(y)$ which is evolutionarily stable. I have been unable to make any progress with this problem analytically. Instead, I consider a discrete version, in which growth stops after a given number of years. This version has the merit of being realistic in many cases, including that of red deer, and of yielding qualitatively interesting results rather easily.

Suppose that typical members of the population start breeding at age $n$, and do not grow subsequently. Let $p_n$ be the probability of surviving from age $n$ to $n+1$; $p_n$ depends on $n$, but not on whether the individual is growing or breeding. Then the probability that an individual will survive to start breeding at age $n$ is

$$S_n = p_0 p_1 p_2 \ldots p_{n-1}.$$

and the expected number of years breeding is

$$Y_n = 1 + p_n + p_n p_{n+1} + \ldots.$$

We now compare the breeding success in a given year of typical members of the population, and of rare mutant individuals which start breeding either at age $n-1$ or age $n+1$. Let $H_n$ = breeding success of typical individuals, and $M_x$ = breeding success of mutants breeding first at age $x$. Then the corresponding fitnesses are

$$W_n = S_n Y_n H_n$$

and

$$\begin{aligned} W_{n+1} &= S_{n+1} Y_{n+1} M_{n+1} \\ &= S_n p_n (1 + p_{n+1} + p_{n+1} p_{n+2} + \ldots) M_{n+1} \\ &= S_n (Y_n - 1) M_{n+1}, \end{aligned}$$

and by a similar calculation,

$$W_{n-1} = S_n (Y_n + 1/p_{n-1}) M_{n-1}.$$

Hence, if a population which first breeds at age $n$ is to be evolutionarily stable,

$$\frac{M_{n+1}}{H_n} < \frac{Y_n}{Y_{n-1}}, \tag{11.2a}$$

and

$$\frac{M_{n-1}}{H_n} < \frac{Y_n}{Y_n + 1/p_{n-1}}. \tag{11.2b}$$

If the first of these inequalities is violated, mutants breeding at age $n+1$ can invade, and if the second is violated, mutants breeding at age $n-1$ can invade.

$M_{n+1}/H_n$ is the proportional breeding advantage of a rare individual which grows until age $n+1$ in a population which stops growing at age $n$. Since the proportional increase in size per year is likely to decrease with age, $M_{n+1}/H_n$ will usually decrease with age $n$. Similarly, $M_{n-1}/H_n$ will usually increase with $n$. If growth ultimately ceases, even in non-breeding individuals, both ratios tend to unity.

Consider first the case with a constant force of mortality (i.e. no senescence), with $p_0 = p_1 = p_2 \ldots p_n = p$, say. For example, suppose

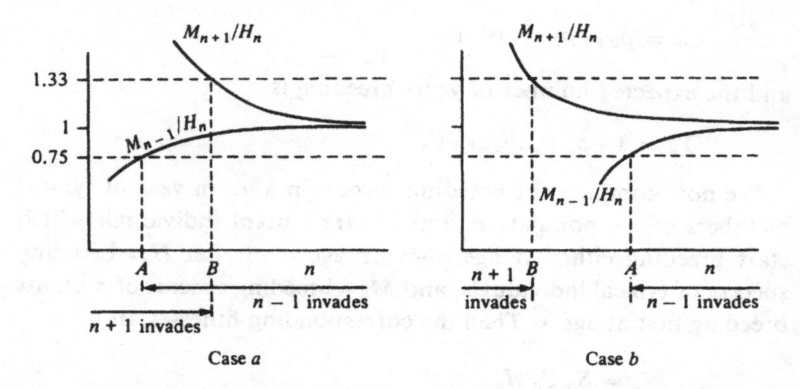


Figure 29. The size game. The upper and lower broken lines are graphs of $Y_n/(Y_n - 1)$ and $Y_n/(Y_n + 1/p_{n-1})$ respectively. For explanation and symbols, see text.

$p = 0.75$. Then $Y_n = 4$, $Y_n/(Y_n - 1) = 1.33$, and $Y_n/(Y_n + 1/p_{n-1}) = 0.75$. There are two cases to consider, illustrated in Figure 29.

In case *a*, there is no single age class which is uninvadable. Any age class in the range *A–B* could be invaded by mutants maturing either earlier or later. The result will be a phenotypically variable population, with phenotypes presumably ranging from *A* to *B*.

In case *b*, any single phenotype in the range *B–A* is uninvadable, by mutants maturing either earlier or later. The result will be a phenotypically uniform population, with its mode in the range *B–A*. The actual mode of the population would depend on its evolutionary history, i.e. on whether the stable range was approached from above or below.

To what biological situations do cases *a* and *b* correspond? In case *a*, the largest individuals in the population gain a substantial advantage ($M_{n+1}/H_n \gg 1$), but the smallest do not suffer a proportional disadvantage (Figure 30*a*). The result is a variable population, and perhaps a dimorphic one. A possible example is the bee *Centris pallida* (see p. 72) in which large males become patrollers and small ones hoverers; the essential point here is that hoverers, although inferior, do not have zero fitness. The variability exists because there is a strategy, alternative to large *n*, which brings adequate returns.

In Figure 30*b*, the smallest members of the population suffer a substantial penalty ($M_{n-1}/H_n \ll 1$), but the largest do not gain a

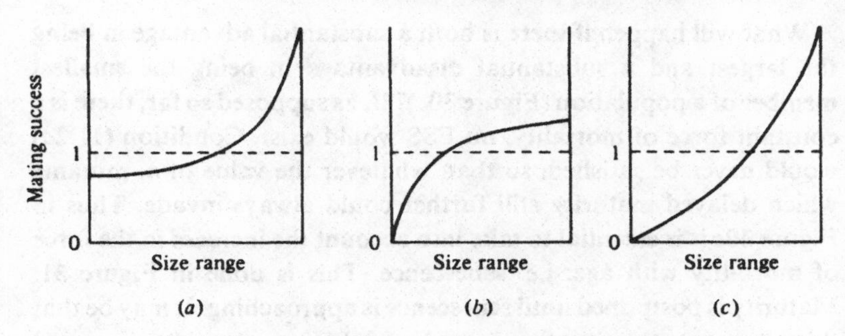


Figure 30. Three ways in which the mating success of an individual may vary with its position in the size range of the population.

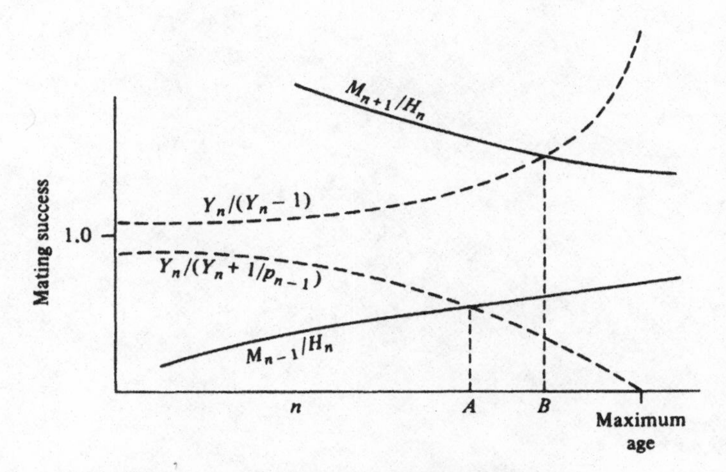


Figure 31. The size game, allowing for senescence. For explanation and symbols, see text.

corresponding advantage. The population is uniform, because neither type of mutant can invade; small mutants have very low fitness, and large ones gain little advantage. A possible example is the wood pigeon: Murton, Westwood & Isaacson (1964) showed that a hierarchy is established in winter flocks, and that individuals at the bottom of the hierarchy often starved.

What will happen if there is both a substantial advantage in being the largest and a substantial disadvantage in being the smallest member of a population (Figure 30*c*)? If, as supposed so far, there is a constant force of mortality, no ESS would exist. Condition (11.2*a*) would never be satisfied, so that, whatever the value of *n*, mutants which delayed maturity still further could always invade. Thus in Figure 30*c* it is essential to take into account the increase in the force of mortality with age: i.e. senescence. This is done in Figure 31. Maturity is postponed until senescence is approaching. It may be that this represents the situation in male red deer.

# 12 *Honesty, bargaining and commitment*

This chapter is concerned with the transfer of information during contests. The process is not well understood from a game-theoretic point of view. A considerable body of data exists, but has in the main been collected by ethologists, who did not appreciate that there are difficulties in understanding why selection should favour the giving of information in such situations. Consequently, the behaviour has often not been related to the contexts in which it occurs in nature, or to its effects on fitness. A proper understanding will be impossible until these cases have been re-analysed in the light of an adequate evolutionary theory.

In the meanwhile, I shall review some of the relevant facts and suggest some possible explanations. In doing so, I have deliberately used human analogies to suggest hypotheses. Of course, students of animal behaviour do this all the time, and I am no exception. The proper procedure, however, is to reformulate the hypothesis, once found, in terms of a population model; if this cannot be done, the hypothesis is inadmissible. I have tried to indicate how this could be done for the ideas discussed here, but the models are worked out in less detail than they should be. My excuse is that a more detailed articulation of the models is unjustified until relevant data are available. This chapter should therefore be treated as speculation about the direction, empirical and theoretical, that research should take.

In considering the evolution of information transfer, there is a basic distinction between two types of information, as discussed on pp. 35–6:

(i) Information about resource-holding power (RHP); i.e. about size, weapons etc. which might influence the outcome of an escalated fight.

(ii) Information about motivation; i.e. about what an animal will do next.

The reason why these two must be distinguished is as follows. A genetic change causing an animal to behave differently in a given situation (i.e. a change in its motivational system) can occur with little selective cost to the animal, except in so far as the change in behaviour itself has selective consequences (e.g. escalation may lead to injury). A genetic change increasing an animal's RHP will be costly outside the contest situation; for example, as discussed in the last chapter, an increase in size of red deer implies delayed reproduction and a greater chance of mortality. Of course, it may be difficult to be sure in actual cases what the selective cost of some change might be; the essential point is that the cost, high or low, must be taken into account in any evolutionary explanation.

No difficulty arises in accounting for the transmission of information about RHP; this is what happens in assessment, as discussed in Chapter 9. The difficulties arise in accounting for information about motivation, essentially because there is nothing to prevent animals 'lying' about what they will do; more formally, it is hard to see how selection could maintain a consistent relationship between signal and subsequent action. The same basic point has been made by Zahavi (1981).

Section A below reviews some of the empirical data. Section B considers the possibility that, since theory suggests that no stable relationship can be maintained between message and meaning, what we observe is a continuously changing state. Sections C and D discuss two processes, 'bargaining' and 'commitment', suggested by analogy with human contests. Bargaining is relevant when resources are divisible; this leads to the suggestion that information transfer has evolved as a necessary component of territorial behaviour. The analysis of commitment leads to the conclusion that there is no close analogy between apparent cases of commitment in animals and the human cases.

## A Information transfer in animal contests

An extensive literature exists to show that, in some sense, information *is* transferred during animal contests (e.g. Stokes, 1962*a*,*b*, Dunham, 1966, Andersson, 1980, on birds; Hazlett, 1966, 1972, Hazlett & Bossert, 1965, Dingle, 1969, on crustaceans; Simpson, 1968, Dow,

Ewing & Sutherland, 1976, Jakobsson, Radesäter & Järvi, 1979, on fish; Rand & Rand, 1976, on lizards; Riechert, 1978, on spiders). These investigations establish four points:

(i) It is common for an animal to use a range of actions during a contest; these actions can plausibly be arranged on a scale of increasing aggressiveness.

(ii) Information is present in these acts, in the sense that there is a correlation between the act now performed and the next act by the same individual.

(iii) Information is received, in the sense that there is a correlation between the act now performed by one individual, and the next act performed by its opponent.

(iv) A common pattern is for the contest to start with acts at a low level on the scale of aggression, and gradually escalate, as each animal matches any increase in aggression by its opponent. Such contests may or may not end in physical contact.

An important question is how far the acts performed by individuals carry information about long-term intentions, and in particular about which will be the ultimate winner. Caryl (1979) has re-analysed the data on tits (Stokes, 1962*a*,*b*), grosbeaks (Dunham, 1966) and skuas (Andersson, 1980) with these questions in mind. His conclusions are as follows:

(i) Displays are poor predictors of physical attack. No one display is followed by attack with high probability. A particular display may be correlated with attack at one time of year but not another. In two out of the three species, the most aggressive display by one bird did not correlate with retreat by the other.

(ii) Some displays are good predictors of immediate retreat.

(iii) Two different displays may have the same effect on the receiver but predict different things about the actor.

The existence of an 'I surrender' signal is easy to understand; similar signals are found in other groups (e.g. Dow *et al.*, 1976, for the fish *Aphysemion striatum*). Apart from this, there is little reason to think that the displays are conveying information about motivation, or about the level to which a bird will escalate.

A rather similar conclusion emerges from the studies of fish. Simpson's (1968) study of *Betta splendens* is particularly illuminating, because he says that 'a criterion for a successful description of the threat display is the prediction of the outcome of encounters from the differences between the displays of the participants'. As explained above, such a criterion seems inappropriate on theoretical grounds; it is therefore interesting that it also fails in practice. Thus, Simpson did find one feature (the proportion of time during which the gill covers are raised) which does predict the outcome, but only during the last two minutes of a fight. Earlier in the same fights, winners exceeded losers in this respect in four cases, losers exceeded winners in two cases, and in another two cases there was no difference. In other words, even when the investigator expected displays to predict outcomes, they failed to do so.

Dow *et al.* (1976), studying *A. striatum*, found that the most aggressive act (biting the opponent) was commoner in eventual winners than in losers in the *first* quarter of a fight, but that subsequently winners and losers were indistinguishable. It seems inconceivable that selection should favour a fish which, at the start of a fight, signals 'I do not intend to continue for long enough to ensure victory', and which then continues without conveying any further information. As the authors acknowledge, the explanation probably lies in the artificial circumstances in which fights took place. Fish were kept in a tank separated by an opaque barrier, and, after a period of familiarisation, fights were initiated by suddenly removing the barrier. The behaviour of a fish immediately after the removal of a barrier might well be inappropriate.

Jakobsson *et al.* (1979) studied fights in the cichlid, *Nannacara anomala*, also in the laboratory. In general, their results resembled Simpson's, in that the slight behavioural differences which did exist between ultimate winners and losers occurred towards the end of a fight. In this study, however, it is known how victory and defeat were determined. The fish were matched carefully for size, differing by less than 0.2 g in a total of 5 g, yet in 10 out of 11 cases the larger fish won. The authors suggest that this accurate assessment of size is made possible by the mouth-wrestling which is a feature of these contests.

It would probably be wrong, though, to conclude that all the displays observed are simply an aid to RHP assessment. Thus Dow *et al.* state that in *Aphysemion* there is no correlation between winning and either size, colouration or fin area. It would also be wrong to conclude that the pattern of display never correlates with winning and losing. As an example, Riechert (1978) has shown that in the spider *Agelenopsis aperta*, when allowance has been made for ownership and size differences, winning spiders show a more varied behaviour than losing ones.

### B Bluff as a transitory phenomenon

Andersson (1980) has offered the following explanation of why many species have a variety of different threat displays. Initially, some movements may be regularly associated with actual attack; in gulls, which Andersson studies, these would include stretching the neck preparatory to pecking, and lifting the carpal joints from the side of the body. So long as such an 'intention movement' is regularly associated with attack, it will be effective in causing an opponent to retreat. It will then pay individuals to use the movement even when their motivation to attack is low. Then, according to the scenario discussed several times already, the signal will lose its effectiveness.

As Andersson points out, however, it does not follow that intention movements will never evolve as signals in the first place, but only that they will not last indefinitely as reliable signals. In fact, one of two things can happen. First, a number of different threat displays may be maintained by frequency-dependent selection. This could happen because the more commonly a particular signal is used as a bluff the less effective it becomes. Alternatively, different threat displays might succeed one another in evolutionary time, as each signal went through the sequence of intention movement to honest signal to bluff to ineffective signal. If this cyclical pattern of evolution is common, we would expect to find that the threat displays of related species differ from one another more than do, for example, alarm notes, which are not used in contest situations.

### C Bargaining, territory and trading

The essential features of a bargaining situation are as follows. Two (or more) contestants are competing for a divisible resource. Each would like a larger share than the other is willing to grant, yet both would prefer to share the resource than to allow the negotiation to break down and an escalated contest to ensue. Territorial contests between animals resemble bargaining in that space is a divisible resource and it will often be the case that two animals would both benefit in fitness terms by sharing the space rather than risking an escalated contest. I shall first discuss the logic of human bargaining situations, and then develop a model of territorial behaviour; for the former topic, I have drawn heavily on the ideas of Reinhard Selten (1975, and personal communication).

Before discussing how people might behave in bargaining situations, two general points must be made about the nature of such contests: first, breakdown must be a possibility and, secondly, we must distinguish games of complete and incomplete information.

If breakdown were not possible, no settlement could ever be reached. Thus suppose two people, *A* and *B*, are debating how a sum of £20 should be divided between them. Each might start by proposing that he receive £19 and his opponent £1, but the other would not agree. If there is no time limit, and no cost to continual negotiation, there is no reason why either should alter his proposal, and hence no way in which the argument can be settled. Breakdown occurs if neither player is prepared to alter his bid. If the result of a breakdown is that neither gets a share of the £20, or that both are involved in an expensive escalated contest, there is a reason for not being too intransigent.

If *A* and *B* are arguing about £20, and both know the sum at stake, this is a game of complete information. Compare this with the following imaginary example of wage bargaining. The management would prefer to give no rise at all, but would pay 10% rather than face a strike. The union would like as big a rise as possible, but would be willing to settle for 5% rather than strike. Clearly, a settlement would be welcomed by both sides at some point between 5 and 10%. The union, however, does not know that the management will go to 10%, and the management does not know that the union would settle for 5%. Further, it would not pay the union to announce right away that it would settle for 5%, because, if it did, that is all it would get, and it is hoping for more. This, then, is a game of incomplete information; each side knows something that the other does not.

The distinction is important for the following reason. In the game of complete information, it seems likely that (in the absence of any asymmetry of power) *A* and *B* will agree to take £10 each. But the union and management cannot agree to split the difference at 7.5%, because neither has the necessary information, and neither will believe what the other says. Consequently, games of incomplete information are more likely to end in breakdown, even when an acceptable compromise was possible.

Before turning to biological problems. a few things can be added about how people can be expected to behave in bargaining situations. These remarks are based partly on theory and partly on observations of actual behaviour. Needless to say, predicted and observed behaviour do not perfectly agree, but there is enough qualitative agreement to justify the following assertions.

(i) Players of bargaining games must have a varied set of possible signals; a negotiator who could only say '20% or we strike' would be unlikely to reach an agreement.

(ii) Breakdowns in negotiation do occur in situations in which it would pay both players to settle; the greater the degree of ignorance each has about the other's situation, the more likely is a breakdown.

(iii) A player who has less to lose from a breakdown is more likely to risk one.

I now turn to the evolution of territorial behaviour. The discussion is unavoidably theoretical. Although much is known about how animals behave once territories have been established, we know little about how those territories are established in the first place. By raising some of the theoretical questions, I hope to encourage people to make the relevant observations.

What would a satisfactory model of territorial behaviour be like? I suggest that it would have the following properties:

(i) A description of how individuals behave in space and time, relative to the position and behaviour of others, sufficiently precise to be programmable on a computer.

(ii) A population of individuals, each with such a behavioural program, should arrange themselves in space in an appropriate manner: e.g. each should own an area adequate for breeding; if too many individuals were initially present in a region, some should be excluded.

(iii) The behavioural program should be evolutionarily stable against plausible mutational changes.

It would be premature to construct such a model before data

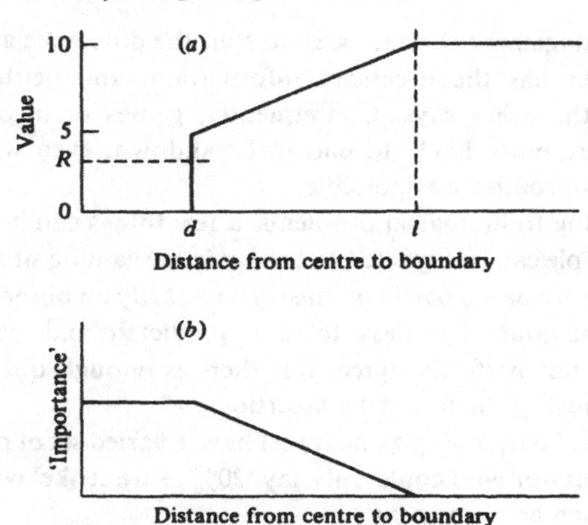


Figure 32. The value of a territory. For explanation, see text.

against which it could be checked are available. It is clear, however, that the behavioural program would have to be subtle and complex. Indeed, my main motive for writing this section is to emphasise the complexity of the problem.

Rather than seek a complete behavioural program, I shall model only the course of a contest between two individuals who have already settled on the positions, in a one-dimensional habitat, of the 'centres' (e.g. nest sites) of their proposed territories. First, some assumptions must be made about the value, in fitness terms, of territories of different sizes; these are shown in Figure 32*a*.

The essential features are as follows. If territory size is less than $d$, it would pay the holder to leave, in which case its expected fitness is $R$ (the fitness if a territory is sought elsewhere). The region within $d$ of the centre will be called the 'central area'. There is no advantage in extending the territory size beyond $l$. Between $d$ and $l$ there is a linear increase in value from 5 to 10. These assumptions are translated into a graph of the 'importance' of particular regions in Figure 32*b*. Note that these assumptions arise from the ecology of the species; they are constraints on the evolution of territorial behaviour, but cannot be changed by that behaviour.

Individuals are capable of displaying with varying intensity, from zero to some maximum level. If two opponents display with maximum intensity at the same spot, this is treated as a breakdown of communication. Individuals then have two choices:

(i) Retreat to edge of own central area, and then, if necessary, escalate.

(ii) Escalate at once without retreating.

If two opponents escalate, the cost to both is $C$. Each has an even chance of 'winning'. The loser must retreat to a distance $l$ from the winner's centre; if this means retreating into the loser's central area, the loser must leave altogether.

I now define three possible strategies:

(i) Hawk. Display to maximum level up to distance $l$. Escalate without retreating if necessary.

(ii) Honest. Behave as Hawk up to distance $d$ only. Between $d$ and $l$ display at a level corresponding to the importance of the place (Figure 32*b*). If opponent signals with higher intensity, retreat, and if at lower intensity, advance.

(iii) Bluffer. Behave as Hawk up to distance $d$ only. Between $d$ and $l$ display with maximum intensity. If opponent also displays with maximum intensity, retreat to edge of own central area.

Other strategies are possible, but these three should be sufficient for our present purpose, which is to see whether the honest use of variable signals can be evolutionarily stable.

It turns out that the ESS depends on the spacing between centres. There are two cases of interest. Case *a*, Figure 33, has the essential feature that if one animal takes the full area it would like, the other must leave. The payoff matrix is shown in Table 25. A few points are needed in explanation:

(i) Any contest involving a Hawk will end with escalation. Both

Table 25. *Payoff matrix for territory game (see Figure 33a)*

| | Hawk | Bluffer | Honest |
|---|---|---|---|
| Hawk | $\frac{1}{2}(10+R)-C$ | $\frac{1}{2}(10+R)-C$ | $\frac{1}{2}(10+R)-C$ |
| Bluffer | $\frac{1}{2}(10+R)-C$ | 7 | 5 |
| Honest | $\frac{1}{2}(10+R)-C$ | 9 | 7 |

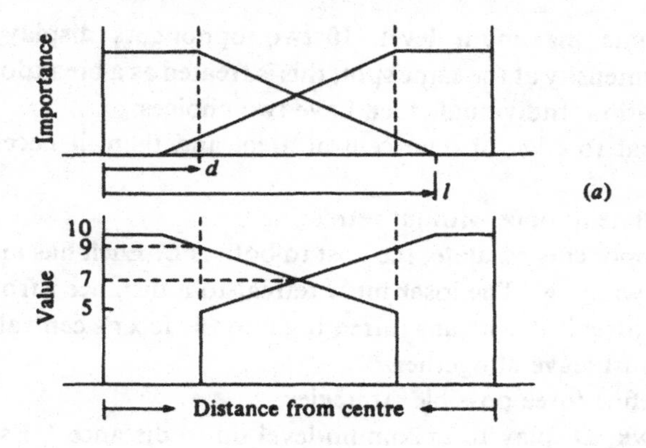


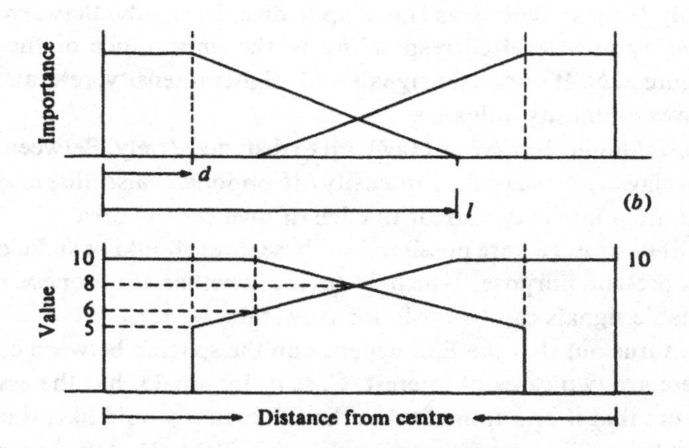


Figure 33. The value of a territory in two cases, differing in the spacing between centres.

contestants lose $C$, and have equal probabilities of gaining a maximum territory $(+10)$ and of leaving $(+R)$.

(ii) Two Honest strategists will divide the space equally.

(iii) Two Bluffers will not escalate, since each will escalate only within its own central area. I have therefore assumed that they will share, and have, at least on average, equal payoffs. But I am not happy about this assumption; in effect, it assumes that if two opponents each discover that the other is bluffing, they revert to honest behaviour.

Table 26. *Payoff matrix for territory game (see Figure* 33a), *assuming* $R=5-2C$

| | Hawk | Bluffer | Honest |
|---|---|---|---|
| Hawk | $7.5-2C$ | $7.5-2C$ | $7.5-2C$ |
| Bluffer | $7.5-2C$ | 7 | 5 |
| Honest | $7.5-2C$ | 9 | 7 |

and, taking $C=1$,

| | Hawk | Bluffer | Honest |
|---|---|---|---|
| Hawk | 5.5 | 5.5 | 5.5 |
| Bluffer | 5.5 | 7 | 5 |
| Honest | 5.5 | 9 | 7 |

(iv) Honest v. Bluffer. The Bluffer is revealed as such when it advances into Honest's central area. Bluffer then retreats to the edge of its own central area, and remains there. The essential assumption is that if, in such a contest, Bluffer is revealed as such, Honest gets a larger share than Bluffer. It does not matter how much larger, the same conclusions would follow.

We can simplify the matrix by recognising a necessary relationship between $R$ and $C$. Thus at the edge of the central area the payoffs for escalating and for leaving must be equal; it would not be worth escalating to win a territory less than $d$, and it would not be worth leaving if it were possible to win a territory larger than $d$.

Hence $\frac{1}{2}\times 5+\frac{1}{2}\times R-C = R$

or $R = 5-2C$.

The resulting payoff matrix is shown in Table 26. Thus, provided the cost of escalation is not too small, the ESS is Honest. Escalation will not occur, and honest graded signals will be given.

I now consider the case illustrated by Figure 33*b*, in which, even if one bird occupies its maximum desired territory, there is still room for the other to have more than $d$. With the same assumptions as before, the payoff matrix is shown in Table 27. Now Honest will never be a component of the ESS. If $C<2$, Hawk is the ESS; if $C>2$, there will be a mixed Hawk/Bluffer ESS.

In concluding that Honest is the ESS for case *a*, and Hawk or Hawk/Bluffer for case *b*, it has been tacitly assumed that this is a game

Table 27. *Payoff matrix for territory game (see Figure* 33b)

| | Hawk | Bluffer | Honest |
|---|---|---|---|
| Hawk | $8-C$ | 10 | 10 |
| Bluffer | 6 | 8 | 10 |
| Honest | 6 | 6 | 8 |

of complete information. That is, an animal knows it is playing a game of, say, type *a*, or, more precisely, all contests are of type *a*, so that an appropriate behaviour can evolve. In practice, the game might be one of incomplete information. If some contests are of one type and some of another, then two situations are possible:

(i) All an individual knows is how its opponent is behaving. The game is one of incomplete information. The ESS will depend on the relative frequencies with which the two types of game occur. Like human games of incomplete information, however, it seems that escalation will sometimes occur when it would have paid both contestants to settle.

(ii) Each bird knows the location of its opponent's centre, and hence knows the type of contest it is engaged in on any particular occasion. This seems unlikely, but, if it was true, each contest would be one of complete information. It is therefore interesting that, although some escalation would occur in type *b* contests, it would never happen that an individual would be forced to leave when compromise was possible. It is also interesting that behaviour would be more aggressive in the less dense population, which is contrary to what intuition might suggest. The reason is that individuals have more to lose in type *a* contests, because losing an escalated fight means leaving the area altogether, whereas in type *b* it means only getting a smaller territory.

This model is a complicated one; the reader will probably feel it is too complicated. Yet, on the one hand, I am unable to propose a simpler model which can give any account of the evolution of territorial behaviour at all, and, on the other, I have made a number of unrealistically simple assumptions (e.g. a linear habitat, territory centres already fixed). The following conclusions do emerge:

(i) There are situations in which the giving of graded signals which honestly convey information about the value of a region to the contestant can be evolutionarily stable.

(ii) Contests can result in compromise, but only if breakdown and escalated fighting is a possibility.

(iii) If information is incomplete, escalation may occur even when compromise would have paid both contestants better.

I know of no data on territorial behaviour which could be used to make the preceding model more realistic. Perhaps the best example of bargaining behaviour occurs in the very different context of 'egg-trading' in the coral reef fish the black hamlet, *Hypoplectus nigricans*, described by Fischer (1980). These fish are simultaneous hermaphrodites with external fertilisation. Eggs are planktonic, so there is no parental care. The fish are self-fertile, but in practice self-fertilisation is prevented because eggs and sperm are shed at different times. Mating occurs in pairs, each fish fertilising the eggs laid by its partner.

The fish are active only during the day. Spawning occurs during the last two hours before sunset. During this period, a pair of fish will come together and engage in a 'spawning bout'. Most fish engage in only one such bout in a single evening, but a minority engage in two, and occasionally in three, bouts. During a bout, a number of spawning acts occur. Each act is initiated by the partner which will play the female role. Almost always, spawning is immediately preceded by a particular display, the 'head snap', and egg-laying is always accompanied by a characteristic quivering of the body. During a typical bout with the same partner, a fish will engage in 4 to 12 spawning acts, about equally frequently as a male and as a female. Male and female roles alternate rather regularly during a bout. It is this alternate laying of eggs which is described as egg-trading.

To understand what is happening, we must ask what behaviour is to be expected from a simultaneous hermaphrodite. If, as is likely (and has been checked for hamlets), eggs are more expensive in resources than sperm, a hermaphrodite can increase its fitness by fertilising the eggs of several other individuals, while ensuring that its own eggs are fertilised. Thus suppose that, in a single spawning bout, first one partner laid all its eggs, and then the other. It would then pay each of these fish to pair with a fresh partner and to fertilise its eggs also; the new partner would be the loser, because it would have no eggs to fertilise. Also, since hamlets do not have mature eggs to lay every day, but could easily produce sperm every day, it would also pay fish to pair on days when they had no eggs. Fischer suggests that egg-trading evolved to prevent this kind of cheating. Thus suppose fish *A* lays only a few eggs on its first spawn. Then if *A*'s partner does not reciprocate by laying eggs in turn, *A* can leave and seek a new partner, while still in possession of eggs to use in trading with the new partner.

There is some evidence to support this interpretation. First, the length of time between one act of egg-laying by a fish and the next is shorter if the first act is reciprocated by the partner laying eggs. Secondly, fish laid fewer batches of eggs in those (relatively rare) bouts in which their partner laid no eggs. Thirdly, fish which paired with more than one partner during an evening were not more successful (as judged by total number of spawning acts engaged in, as either male or female) than fish pairing only once.

These facts, together with the initiative taken by the fish occupying the female role at any time, and the characteristic quiver associated with egg-laying, are all consistent with the suggestion that egg-trading has evolved as a device which prevents fish from being exploited by partners which have no eggs to lay. It is not known whether fish can detect the laying of eggs by their partner; if they cannot, then the system would not be stable against a cheating mutant which quivered without laying eggs. But provided that there is a necessary connection between quivering and egg-laying, as in practice there appears to be, then there is little to be gained by a fish which pairs when it has no eggs.

The system is further stabilised by the fact that mating is confined to a few hours before sunset. A fish which leaves its partner may not have time to complete a spawning bout with a second one; it is usually some 20 minutes after the initiation of pairing before one fish lays eggs.

The 'egg-trading game' does illustrate some features of bargaining games. It is, at least potentially, a game of incomplete information; each fish knows whether it has eggs to trade, but not whether its partner has eggs. The two partners have a common interest in completing a spawning bout, but there is a potential advantage to each of them in not reciprocating. There is a time-limit, dusk, before which negotiations must be completed. The strategy which has evolved is a graded one, of laying eggs in small batches rather than all at once.

A striking example of trading has been described by Hazlett (1978, 1980) in hermit crabs. There is good evidence that empty shells are a limiting resource for these crabs. An individual may find itself in a shell which is either too large or too small. A large crab in a small shell and a small crab in a large shell can both benefit by an exchange, and such exchanges do in fact take place. One crab will initiate an exchange by tapping or shaking the shell of another in a manner which is characteristic of the species. The non-initiating crab may stay inside its shell, or it may withdraw from its shell after first signalling by tapping the initiator on its chelipeds; in the latter case, an exchange of shells occurs.

If an exchange of shells would leave the non-initiating crab in a shell further from its preferred size, then in a majority of cases no exchange takes place. Neither the size nor the sex of the initiating crab influences the likelihood of an exchange, so it seems that an exchange requires mutual benefit, and cannot be enforced by the initiator. Exchanges take place between members of different species, provided that the initiator has in its repertoire a signal appropriate to the other species. Since different species of crab prefer different shell species, interspecific exchanges afford opportunities of mutual benefit additional to those arising in intraspecific interactions.

In this example, no special disadvantage is associated with a failure to agree. Consequently no logical difficulties arise in explaining the exchange of information. It pays each crab to acquire information about whether an exchange would benefit it.

## D Commitment

In some animal contests, individuals signal their intentions beforehand. For example, in the Harris sparrow, to be dark signals aggression (p. 82). An African elephant in musth (an aggressive state into which adult males enter periodically) signals aggression by a visible and odorous secretion and by urinating (Poole & Moss, 1981). This suggests an analogy with human contests in which prior commitment can be advantageous. I will first discuss the logic of commitment in human contests; the matter is discussed in more detail by Hirschleifer (1980).

Table 28. *The Prisoner's Dilemma*

| | Defect | Cooperate |
|---|---|---|
| Defect | 0, 0 | 2, −1 |
| Cooperate | −1, 2 | 1, 1 |

First, consider commitment in the bargaining games discussed. Suppose that *A* and *B* are debating the division of £20, but that *A* has the privilege of making the first bid, and suppose also that this bid, once made, is unalterable. It would then pay him to say 'I demand £19; you can have £1'. *B* would then have the choice of accepting £1, or refusing, in which case the negotiation would break down and he would get nothing. Logically, *B* should accept the £1; hence prior commitment enables *A* to get £19 instead of only £10. In practice, it might pay *A* to be a little less greedy, for fear that *B* might refuse out of pique.

As expected, things are a little less simple in games of incomplete information. For example, suppose that, in the wage-bargaining case, the union side could commit itself – perhaps by a vote of the membership – to strike unless it got 10%. The management would have to agree, and the union would have got the maximum rise possible. The snag, of course, is that only the management knows that 10% is its upper limit. If the union miscalculated, and committed itself to strike unless it got 15%, there would be a strike, although the union would have preferred to settle.

The same idea can be applied to the Hawk–Dove game, and to the classic Prisoner's Dilemma game (see Appendix K), shown in Table 28. The latter is a symmetric game in which it pays a player to defect, whatever his opponent does, so the ESS is to defect; yet both players would be better off if they cooperate. We suppose that one player, *A*, is privileged to commit himself to a course of action, and that the

Table 29. *Commitment in the Hawk–Dove and Prisoner's Dilemma games*

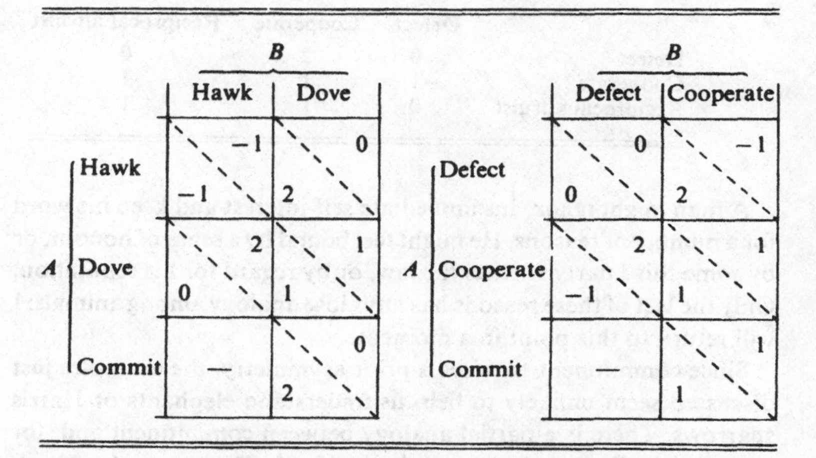

| | | B | |
|---|---|---|---|
| | | Hawk | Dove |
| A | Hawk | −1, −1 | 2, 0 |
| | Dove | 0, 2 | 1, 1 |
| | Commit | — | 2, 0 |

| | | B | |
|---|---|---|---|
| | | Defect | Cooperate |
| A | Defect | 0, 0 | 2, −1 |
| | Cooperate | −1, 2 | 1, 1 |
| | Commit | | 1, 1 |

other, *B*, does whatever is best in the light of *A*'s choice. The results are shown in Table 29.

In the Hawk–Dove game, the best policy for *A* is to say 'whatever you do, I shall choose Hawk'; the best *B* can do in reply is to choose Dove. In the Prisoner's Dilemma, it would not pay *A* to commit himself to defect, because then *B* would also defect, and *A* would get 0. *A* should say 'If you cooperate, I will cooperate; if you defect, I shall defect'. Then *B* will cooperate, and *A* will get 1.

Two points need emphasising:

(i) Commitment, in these examples, requires a prior asymmetry. One contestant is in a privileged position. It would be no use if both demanded £19, or were both committed to choosing Hawk.

(ii) The commitment must be announced, and, once announced, it must be irrevocable. There is, however, a significant difference between the Hawk–Dove and Prisoner's Dilemma games. In the former, *A* commits himself to Hawk; once *B* has chosen Dove, *A* would not wish to break his promise, because Hawk is the best reply to Dove. In the Prisoner's Dilemma, however, *A* makes a commitment which depends on *B*'s choice; once *B* has chosen to cooperate, it would pay *A* to switch to defect, but he is prevented from doing so by his prior commitment.

Table 30. *Payoff matrix for the repeated Prisoner's Dilemma*

| | Defect | Cooperate | Reciprocal altruist |
|---|---|---|---|
| Defect | 0 | 2 | 0 |
| Cooperate | −1 | 1 | 1 |
| Reciprocal altruist | 0 | 1 | 1 |

A man might ignore his immediate self-interest and keep his word for a number of reasons. He might feel bound by a sense of honour, or by some third party such as the law, or by regard for his reputation. Only the last of these reasons has any close analogy among animals; I will return to this point in a moment.

Since commitment requires a prior asymmetry, the examples just discussed seem unlikely to help us understand elephants or Harris sparrows. There is a partial analogy between commitment and, for example, the Bourgeois strategy in the Hawk–Dove game (p. 22). A Bourgeois strategist can be seen as an individual genetically committed to choosing Hawk when owner; announcement is not needed because ownership itself amounts to an announcement. However, we must distinguish between the Hawk–Dove game and the Prisoner's dilemma; a Bourgeois strategy is a solution only for the former.

Thus consider the asymmetric Prisoner's Dilemma. One player, *A*, can make a prior commitment. The strategy 'If *A*, promise to cooperate if one's opponent cooperates, and to defect if he defects', is *not* an ESS if breaking the promise is possible. A commitment binding *A* to do something not in his short-term interests is unstable unless it can be enforced. In man, it can be enforced by regard for reputation. It does not pay to go back on one's word, because next time one will not be believed. The analogue of this in animals is Trivers' (1971) concept of reciprocal altruism. Thus suppose an individual plays the Prisoner's Dilemma repeatedly against the same opponent. A reciprocal altruist starts with the choice 'cooperate', and in subsequent games, continues to cooperate against opponents who have cooperated, and to defect against those who have defected. If we assume a long sequence of games against each opponent, so that the payoff in the first game is a negligible part of the total payoff, we get the payoff matrix shown in Table 30. Reciprocal altruist is the only

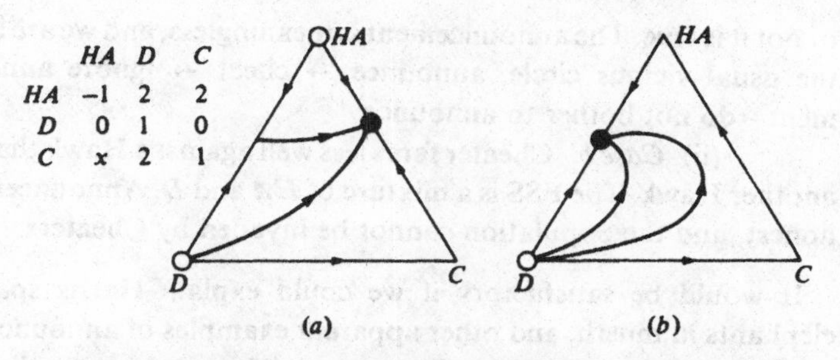

| | *HA* | *D* | *C* |
|---|---|---|---|
| *HA* | −1 | 2 | 2 |
| *D* | 0 | 1 | 0 |
| *C* | *x* | 2 | 1 |



**Figure 34. The Announcer–Dove–Cheater game (*a*) $x > -1$; (*b*) $x < -1$.**

ESS of this game. I return to the problem of the repeated Prisoner's Dilemma in the next chapter.

The two illustrative examples, elephants and Harris sparrows, can reasonably be interpreted as cases of the Hawk–Dove game in which the population has reached a mixed ESS. In the Harris sparrow, some individuals are always Hawks and others Doves (at least for a season), whereas in elephants the same individual can switch from one strategy to the other. The problem is to explain why the Hawk strategists announce that they are Hawks? The announcement may well save time and energy, provided it is believed; thus the problem reduces to explaining why cheating does not spread.

To answer this question, consider the game illustrated in Figure 34. There are three pure strategies:

(i) *HA*, Hawk–Announcer. Always choose Hawk, and announce that one will do so.

(ii) *D*, Dove. Always choose Dove.

(iii) *C*, Cheater. Announce that one will choose Hawk, but retreat if opponent escalates.

In drawing up the payoff matrix, I have assumed that Cheater successfully frightens off Dove, and that two Cheaters can divide the resource (or have equal chances of obtaining it). The crucial question is how Cheater fares against Hawk. There are two possibilities:

(i) *Case a.* Cheater does avoid dangerous escalated contests against Hawks. The ESS is a mixture of *HA* and *C*. That is, all members of the population announce that they are Hawks, whether or not it is true. The announcement is meaningless, and we are back in the usual vicious circle, announce → cheat → ignore announcement→do not bother to announce.

(ii) *Case b.* Cheater fares less well against a Hawk than does another Hawk. The ESS is a mixture of *HA* and *D*. Announcement is honest, and the population cannot be invaded by Cheaters.

It would be satisfactory if we could explain Harris sparrows, elephants in musth, and other apparent examples of announcement, as corresponding to case *b*. The essential feature is that a dishonest announcement of aggressiveness is less profitable than genuine aggression. Rohwer's results on Harris sparrows (p. 83) suggest that this is indeed the case. A subordinate bird painted dark is driven out by dark birds, presumably because one dark bird does not want to share with others its opportunity of exploiting pale birds.

The elephant case is harder to understand. It may well be a mixed ESS: so long as musth is rare, a male in musth is unlikely to meet another in the same state, and so gains the advantage of more matings without too high a risk of escalated fighting. The difficulty is to explain why lying does not spread. It may be that the correct explanation is the rather boring one, that it has been impossible to separate the aggressive behavioural consequences of an animal's hormonal state from the overt signalling of that state.

In conclusion, there is no close analogy between commitment as a strategy in human contexts, and cases in animals in which a particular behaviour is signalled beforehand. The human examples require a prior asymmetry, so that one contestant only is in a position to commit himself. The analogy, therefore, is to asymmetric animal contests. As a solution to the Prisoner's Dilemma paradox, commitment must be binding, leading individuals to do things not in their short-term interests. This suggests that there is a third party able to enforce the commitment, for which there is no obvious animal analogy, or that contests are not isolated events, so that a reputation for honesty is worth acquiring; in this case, reciprocal altruism is the animal analogy. The apparent examples of 'announcement' in animals may depend either on some special circumstance which makes lying unprofitable, or on a physiological constraint making it impossible.

# 13 *The evolution of cooperation*

The evolution of cooperation among animals, either within or between species, presents an obvious problem for Darwinists. Darwin himself recognised this. He thought that social behaviour, as shown by social insects, could be explained by family selection. He also remarked that his theory would be disproved if it could be shown that some property of one species existed solely to ensure the survival of another; by implication, mutualism between species must be explained by indirect benefits to the individuals performing cooperative acts.

Despite these essentially correct insights, little progress was made in analysing the selective forces responsible for the evolution of cooperative behaviour for 100 years after the publication of *The Origin of Species.* For many biologists, it was enough that a trait could be seen to favour the survival of the species, or even of the ecosystem. It is clear from the writings of J.B.S. Haldane and R.A. Fisher that they knew better than this. The decisive turn to the significance of kinship in social behaviour, however, was by Hamilton (1964). Perhaps inevitably, the elegance of the idea of inclusive fitness and the prevalence of genetic relationship between members of social groups tended to distract attention from other selective forces, and in particular from mutualistic effects; that is, from the fact that two (or more) individuals may cooperate because it benefits both to do so. However, the significance of mutualistic effect was not lost sight of (e.g. Michener, 1974; West-Eberhard, 1975), and Trivers (1971) pointed out the possibility of delayed reciprocation.

Today it is clear that the evolution of social behaviour has involved both interactions between kin, and mutual benefits to cooperating individuals. In this chapter, I want to approach the problem from a different direction, although the conclusions reached will be similar. I start by describing the work of Axelrod (1981) on the evolution of cooperation in the Prisoner's Dilemma game. Then, following Axelrod & Hamilton (1981), I discuss how Axelrod's ideas might be relevant to animal evolution. Finally, I turn to the problem of human cooperation, and the possible relevance of evolutionary game theory to cultural evolution.

The Prisoner's Dilemma is a symmetric two-person game, in which the alternative plays are 'Cooperate' and 'Defect'. The payoffs are shown in Table 28, p. 162. Note that if player *B* cooperates, it pays *A* to defect; also, if player *B* defects, it pays *A* to defect. That is, it pays *A* to defect no matter what *B* does; similarly, it pays *B* to defect no matter what *A* does. Yet, if both defect, they do less well than if both cooperate; this is the paradox.

Considered in an evolutionary context, if each individual plays only a single game against each opponent, the only ESS is to defect. Cooperation will not evolve. Things are different, however, if individuals play repeatedly against the same opponent; it is this situation which Axelrod has studied. He invited a number of people to contribute a computer program to play the Prisoner's Dilemma repeatedly against the same opponent. He then ran a tournament between the programs submitted, each program playing 200 games against each other one. The programs were then ranked according to the total payoff accumulated (not, it should be noted, according to the number of opponents defeated in the individual matches). The winning program, submitted by Anatol Rapoport, was also the simplest. It was 'TIT FOR TAT'; it chooses Cooperate on the first move, and on all subsequent moves makes the choice adopted by its opponent on the previous move.

The results of this tournament were published, and people were invited to submit programs for a second tournament. This was identical in form to the first, except that matches were not of exactly 200 games, but were of random length with median 200; this avoids the complication that programs may have special rules for the last game. Again, TIT FOR TAT was the winner. Further computer analysis showed that if a succession of tournaments were played, with programs increasing in representation if they did well, then TIT FOR TAT ultimately displaced all others. That is, for the programs submitted, TIT FOR TAT was an ESS.

What properties make TIT FOR TAT an ESS? Axelrod suggests that a successful strategy must be 'nice', 'provokable' and 'forgiving'. A nice program is one which is never the first to defect. In a match between two nice programs, both do well. A provokable strategy responds by defecting at once in response to defect; a program which does not at once respond in this way encourages its opponent to defect. A forgiving strategy is one which readily returns to cooperation if its opponent does so; unforgiving strategies are likely to get involved in prolonged periods in which both defect.

Axelrod has later been able to prove that TIT FOR TAT is stable against invasion by any other possible strategy, provided that the sequence of games against each opponent is long enough; his proof is summarised in Appendix K. However, TIT FOR TAT is not the only ESS. Thus 'Always defect' is an ESS. A comparison of *TFT* = TIT FOR TAT and *D* = Always defect shows that the former has a much wider basin of attraction. Thus suppose that individuals play mainly against their neighbours, and that there is some clustering of individuals with similar strategies. It then turns out that *TFT*, even when rare, can invade *D*, because *TFT* does well against neighbours with the same strategy; clustering does not enable *D* to invade *TFT*. This illustrates the fact that kin selection is often needed for the initial spread of a trait which, once it is common, can be maintained by mutualistic effects alone.

These results provide a model for the evolution of cooperative behaviour. At first sight, it might seem that the model is relevant only to higher animals which can distinguish between their various opponents. Thus if such distinction was not possible, an individual which met defection from one opponent would defect against others, and the result would soon be general defection. Axelrod & Hamilton (1981) point out, however, that the model can be applied if each individual has only one opponent in its lifetime. With this proviso, TIT FOR TAT can evolve as a strategy in completely undiscriminating organisms. They also point out that the model can be applied to interactions between members of different species.

If cooperation, within or between species, is to evolve by this road, there are three requirements:

(i) There must be repeated interactions between the same pair of individuals (or, conceivably, between two clones or two endogamous groups).

(ii) Each partner must be able to retaliate against defection by the other.

(iii) *Either* individual recognition must be possible, *or* the number of potential partners with whom an individual interacts must be small, preferably only one.

Axelrod & Hamilton discuss various examples of cooperation and its breakdown from this point of view. First, many examples of mutualism involve interactions between one member of one species and one of another, as for example in sea anemone and hermit crab, or tree and mycorrhizal fungus. Sometimes the number of interacting partners is effectively limited by interacting at a particular site, as in the cleaning fish discussed by Trivers (1971).

The possibility that cooperation may evolve, not between individuals, but between endogamous groups has been discussed by D.S. Wilson (1980; for a mathematical treatment, see Slatkin & Wilson, 1979) under the term 'indirect effects'; although he does not explicitly refer to endogamy, it is clear that the mechanism he has in mind requires a high degree of population viscosity between generations. As an example, Axelrod & Hamilton point to the fact that ant colonies participate in many symbioses, whereas honey bees, which move from place to place more frequently, have many parasites but no known symbionts.

In higher organisms, cooperation can depend on individual recognition. Perhaps the clearest example is Packer's (1977*b*) demonstration of reciprocal altruism between pairs of male olive baboons, in the absence of any known genetic relationship between the interactants. As argued on p. 164, reciprocal altruism of this kind can readily be modelled as a game, with TIT FOR TAT as the ESS.

Applying these ideas to baboons, and *a fortiori* to men, raises questions about the nature of the 'hereditary' mechanism – genetic or cultural – underlying the evolutionary process. Thus the conclusion that cooperative behaviour is a stable outcome rests on the assumption that individuals who are in some sense successful pass their characteristics on to more 'descendants' than those who are not. Three hereditary mechanisms are conceivable:

(i) *Genetic*. The assumption is that differences between individuals adopting different strategies are, at least in part, genetic: i.e. caused by differences between the fertilised eggs from which they developed. It seems to me important that terms such as 'genetically determined' or 'innate' should be used in this rigorous sense. This has by no means always been the case in discussions of sociobiology. For example, E.O. Wilson (1978) starts the chapter on aggression as follows: 'Are human beings innately aggressive? This is a favourite question of college seminars and cocktail party conversations, and one that raises emotion in political ideologues of all stripes. The answer to it is yes.' Yet it turns out, on reading the rest of the chapter, that the only proposition Wilson defends is that human beings are sometimes aggressive.

It seems unlikely that the reciprocal altruism in baboons, discussed by Packer, is maintained solely by selection acting on genetic differences. Some degree of imitation, and perhaps insight learning (i.e. a calculation of the costs of not reciprocating), are probably also involved. In man, such processes are likely to predominate. Nevertheless, genetic evolution may have made both baboons and people readier to learn some things than others.

(ii) *Learning*. In Chapter 5, it was shown that a generalised learning rule, not specific to any particular game or problem, can take a population to an ESS in one generation. But such a learning rule cannot help much in the present context. Thus the type of learning rule which was discussed had, as its starting point, a set of possible 'behaviours'. In the context of a repeated Prisoner's Dilemma, this would require that an individual start with a set of possible strategies, of which TIT FOR TAT would be one; that he play a succession of long matches against individual opponents, adopting different strategies for different matches; and that he gradually adjust the frequencies of the different strategies in accordance with outcomes. One lifetime would not be long enough for such an inefficient learning process.

For man, one can consider the alternative possibility of insight learning. A man might imagine a series of matches, adopting different strategies, and thereby calculate that TIT FOR TAT was best. This also seems implausible; the scientists who participated in Axelrod's tournaments were apparently unable to perform this calculation.

Learning would, however, be important in maintaining TIT FOR TAT, once established. Thus if almost everyone is playing TIT FOR TAT, it would not pay an individual to adopt any other strategy. This could readily be learnt by trial, either in practice or in imagination.

(iii) *Cultural inheritance*. Suppose, to take an oversimplified model, that individuals acquire their behaviour by learning or imitation from others, and that they are more likely to copy successful mentors. Such a process of cultural inheritance would lead to the spread of behaviour patterns, including cooperation, which meet the criteria of evolutionary stability. Such processes of cultural inheritance, and their interaction with genetic processes, have been discussed in much greater depth by Feldman & Cavalli-Sforza (1976) and Lumsden & Wilson (1981). My only purpose here is to raise the question of what types of cultural heredity would be formally similar to asexual genetic inheritance, in the sense of leading to evolutionarily stable states.

I am, in fact, more interested in raising this question than in solving it. Game-theoretic ideas originated within sociology. Naturally enough, the solution concepts which developed were based on the idea of rational calculation. The ideas were borrowed by evolutionary biologists, who introduced a new concept of a solution, based on selection and heredity operating in a population. If, as seems likely, the idea of evolutionary stability is now to be reintroduced into sociology, it is crucial that this should be done only when a suitable mechanism of cultural heredity exists.

It may be that cultural processes will often mimic genetic ones; but there is one distinction which needs to be made between kinds of cultural inheritance. First, consider a case in which all children acquire some trait by imitating their mothers, and in which mothers pass on the trait which they themselves acquired. In the evolution of such traits, 'fitness' would be measured by the Darwinian fitness of mothers; those traits would increase which enabled their possessors to survive and have more children. At the opposite extreme, suppose that each child acquires some trait by imitating a mentor who is not a parent, but who is judged to be 'successful' by some criterion. Traits will then increase which ensure 'success', however that is measured. Since the criteria of success are themselves to some degree culturally determined, a much more complex, but perhaps more realistic, process is involved.

I have written so far as if behavioural traits are properties of individuals, and as if individuals acquired a behaviour for life. Neither of these things is true. Individuals may change their behaviour, and horizontal as well as vertical transmission can occur. Some customs and practices may be properties of institutions – firms, schools, regiments etc. – and at least some such institutions may grow at rates determined by their practices.

In conclusion, there is at least one kind of game which people play, but which seems beyond the capacity of animals. This is the 'social contract' game. Thus suppose that some pattern of behaviour – for example theft – is seen to be undesirable. A group of individuals capable of symbolic communication can agree not to steal, and to punish any member of the group who does steal. That, by itself, is not sufficient to guarantee stability, because the act of punishing is presumably costly, and therefore individuals would be tempted to accept the benefits of the contract but not the costs of enforcing it. Stability requires that refusal by an individual to participate in enforcing the contract should also be regarded as a breach which will be punished. At a later stage, enforcement is entrusted to a subgroup, who are rewarded for carrying it out.

# 14 *Postscript*

To conclude, I will attempt to say where evolutionary game theory came from, what is its present state, and how it may develop in the future.

In origin, the concept of an ESS is highly polyphyletic. For me, it arose from an attempt to formalise some ideas of the late Dr George Price about animal contests: hence the term 'strategy', and a model based on pairwise contests. The idea that variable behaviour could be explained by frequency-dependent selection was proposed independently by Gadgil (1972), and Parker (1970*a*) had interpreted dung fly behaviour in terms of fitness equalisation. A very similar idea had arisen in the study of the evolution of the sex ratio. The original concept is Fisher's (1930). Shaw & Mohler (1953) sought an equilibrium by equalising the payoffs for producing sons and daughters, and MacArthur (1965) used techniques very similar to those in this book to find the stable sex ratio. Hamilton (1967), tackling the problem of the sex ratio in a structured population, explicitly used game-theoretic ideas; his 'unbeatable strategy' is essentially equivalent to an ESS. Fretwell & Lucas (1970), in their analysis of animal dispersal, used the fact that fitnesses must be equal at an equilibrium to reach the concept of an ideal free distribution. More recently, it has become apparent that Trivers' (1971) idea of reciprocal altruism is best understood as the ESS for a repeated game between the same two opponents, and Axelrod & Hamilton (1981) have taken a step towards uniting classical and evolutionary game theory by applying evolutionary concepts to the Prisoner's Dilemma.

It has been exciting gradually to appreciate the way in which these various lines of thought have converged, and to recognise the wide range of applications of game-theoretic ideas in evolutionary biology. This multiple origin, however, has also caused some confusions. Of these, the most important have probably been:

(i) The failure to distinguish between a population of genetically identical individuals adopting a mixed ESS, and a genetically polymorphic population in an evolutionarily stable state.

(ii) The failure to distinguish between the case in which each individual plays one or a series of pairwise contests, leading to conditions (2.4*a*,*b*) for uninvadability, and that in which an individual's fitness depends on some property of the population as a whole, leading to conditions (2.9).

In this book, I have tried to remove some of these confusions. The attitude I have taken is to define an ESS (p. 10) in terms of its uninvadability, and to treat conditions (2.4*a*,*b*) and (2.9) as the criteria of uninvadability for specific models. I think that most future applications are likely to be to 'games against the field' (p. 23), and hence that Hammerstein's proposed conditions (2.9) will be more widely useful than (2.4*a*,*b*).

I now discuss the following in turn:

(i) The field of application of game theory, its limitations, and some specific problems which have proved difficult to treat.

(ii) Areas of biology where game-theoretic ideas seem likely to be particularly fruitful.

(iii) Theoretical problems in evolutionary game theory.

(iv) The relation between game theory and population genetics.

As stated at the outset, evolutionary game theory is a way of thinking about the evolution of phenotypes when fitnesses are frequency-dependent. It is much less useful for analysing the causes of genetic variability. Often the major difficulty in applying the method is in deciding what is the appropriate 'phenotype set' (equivalently, in deciding what are the 'developmental constraints', or in choosing the 'trade-off functions'). Obviously one cannot decide which is the best phenotype unless one knows what are the possible phenotypes. Ideally, the phenotype set could be discovered by studying the range of intraspecific variability under intense selection over a long time-scale; in practice, we will often have to rely on a mixture of common sense and a study of intra- and inter-specific variability. The same difficulty arises in applying optimisation theory (the appropriate alternative to game theory when fitnesses are independent of frequency). There is no way of avoiding it if we are going to understand the evolution of particular adaptations.

One particular point about plausible phenotype sets is worth making. It often turns out that to find the ESS, or to find an optimal strategy, requires formidable mathematical efforts. It would, however, be naïve to suppose that the animals themselves are performing complex calculations. Presumably they are following some relatively simple algorithm, which happens to be stable against other equally simple algorithms. It is therefore worth asking, in particular cases, what particular rules of thumb are being followed. That is, in effect, what Harley (1981) did in devising his 'relative payoff sum' (RPS) (p. 59). It is worth emphasising, though, that he did not derive the RPS rule by analysing how animals actually learn, but by seeking a rule which would approximate to the theoretical 'ES learning rule'. The moral of this is that theoretical work (by optimisation or game theory, as is appropriate) can lead to an ideal solution, which may in turn suggest a practicable mechanism which approximates that solution, a point which was made by McFarland (1974) in the context of optimisation.

So far, two contexts have arisen in which the centre of interest is phenotypic evolution, and yet evolutionary game theory has proved hard to apply: these are contests between relatives (e.g. parent–offspring, sibling–sibling), and sexual selection. In the latter case, the difficulty arises because the phenotype of the female (i.e. female preference) is not itself selected, and changes in frequency only because of linkage disequilibrium between genes for female preferences and genes for the male trait. This is a rather special, though fascinating, case, and I see little point in trying to bring it under the aegis of game theory.

Games between relatives raise a more important difficulty, because many intraspecific contests, both in animals and plants, are between relatives. The natural approach is to write down a 'derived inclusive fitness matrix', as in Appendix F, and seek the ESS of that matrix. I do not think this method will lead us far wrong in practice, but there is no guarantee that it will give the correct solution. This is a topic on which further work is needed.

In general, however, provided that contestants are not related to one another, the introduction of sexual reproduction and diploid inheritance makes little difference. Thus suppose we find an ESS on the assumption of asexual inheritance; then, provided that the resulting phenotype can be produced by a genetic homozygote, it will be uninvadable in a sexually reproducing population just as it is in an asexual one.

I now turn to three areas of biology to which game theory may fruitfully be applied:

(i) *Animal behaviour, particularly contest behaviour*. Most of the examples in this book are from this area, so I will make only two points. The first is that there is a need to analyse the genetic and/or developmental mechanisms responsible for variable behaviour, as discussed in Chapter 6. The second is that the significance of information transfer during contests is still very imperfectly understood; I suspect that studies of how territorial boundaries are established would help to clear up the difficulties.

(ii) *Sex ratio and sexual allocation*. I have said little on these topics, because they are treated at length by Charnov (1981). There is, however, one general point worth making. Phenotypic characteristics such as the ratio of male and female progeny produced by a parent, the allocation of resources in a hermaphrodite, or the age of sex reversal, are all concerned with the process of sexual reproduction, but this has proved no barrier to the use of game theory.

(iii) *Growth and life history strategies*. The optimal growth pattern for a plant depends on what nearby plants are doing. A plant growing by itself would not gain, in seed or pollen production, by having a massive woody trunk. Leaves may be selected as much for shading out competitors as for photosynthesis. In other words, functional analysis of plant growth is a problem in game theory, not in optimisation. This is less obviously true for non-sessile organisms, but, as argued in Chapter 11, life history strategies of animals may also have to be analysed as games. Little has as yet been done in this area; exceptions are Mirmirani & Oster (1978), and an unpublished manuscript by T.J. Givnish.

Turning to the theoretical problems requiring investigation, one can distinguish between those thrown up by practical applications, and those arising from previous theoretical work. Of the former kind, one particular problem has been forced on my attention by discussions with Dr Susan Riechert about the spider *Agelenopsis aperta*. This concerns cases in which an animal is engaged in a series of contests. The work of Axelrod (1981) has shown how fruitful it can be to look for an ESS when repeated contests occur between the same two opponents. If successive contests are with different opponents, I have assumed in this book that each contest is independent of previous ones, and that payoffs are additive. This can sometimes be approximately true, but it need not be. For example, if contests are over a long-lasting resource (for example, a web in *Agelenopsis*), then the winner of one contest will be an owner, not an intruder, in the next. Hence, the strategy adopted in one contest affects the role of the contestant in the next. It is not clear to me at present how such cases can be analysed.

I have no doubt that many other theoretical questions will arise in trying to apply evolutionary game theory. As to problems arising within the theory itself, I find it hard to predict, if only because already I feel somewhat out of my depth. It seems, however, that development is likely to be in two directions. One is the analysis of the population dynamics of asexual populations, as indicated in Appendix D. This is likely to be of greater importance in the study of the early evolution of life (Eigen & Schuster, 1977) than of higher organisms, because the assumption of asexual reproduction is justified in the former case, but made only for convenience in the latter. A second direction is more related to classical game theory. In this book I have considered only games which can be represented in matrix form, and games whose strategy set can be defined by one or more continuous variables. More complex games can be represented in extended form, as a tree whose nodes are choice points for one or other contestant. It may prove helpful to use such a representation in an evolutionary context.

The final question concerns the relation between evolutionary game theory and population genetics. For most purposes game theory can proceed without bothering about genetic details; all it need assume is that there is some additive heritability of the trait in question, and that the ESS, pure or mixed, can be produced by a homozygous genotype. There are, however, two contexts in which the disciplines interact:

(i) The payoffs (fitnesses) can only be estimated using explicit genetic models. This difficulty arises in the case of games between relatives, discussed above. It arises in a more tractable form when analysing sex ratio evolution and related problems. In genetic models of sex ratio evolution, the fitnesses are not arbitrary parameters as in most genetic models, but arise necessarily from the phenotype (progeny sex ratio) and the breeding system. In such cases the concept of an ESS is a useful tool within population genetics. Thus the classical approach to such a problem is to consider two (or more) alleles at a locus, and to ascribe phenotypes (e.g. progeny sex ratios) to each of the three genotypes. One can then seek equilibrium gene frequencies, or, with less mathematical difficulty, one can ask whether there is a 'protected polymorphism': the answer is yes if one allele, *a*, can invade an *AA* population, and *A* can invade an *aa* population.

The ESS approach (Eshel, 1975; Charlesworth, 1977) is to ascribe an arbitrary phenotype, *s*, to a genotype, *AA*, and to ask whether there is any value of *s* such that no mutant with a different phenotype can invade. The method still rests on population genetics, but for some purposes is both simpler and more powerful.

(ii) Game theory predicts a mixed ESS, but no homozygous genotype produces that phenotype; in the extreme case, each genotype may specify a pure strategy. In such cases, it is natural to ask whether there will be a stable genetic polymorphism, with morph frequencies equal to the ESS frequencies. This will not always be the case, even if inheritance is asexual (Appendix D); but often it will be so (pp. 40–3). It is important to find out what conditions must hold if there is to be an equivalence between mixed ESS and evolutionarily stable state; Eshel (1981*b*) takes this question further than I have done here.

# Appendixes

## A Matrix notation for game theory

Throughout this book I have used the notation $E(p,q)$ to express the payoff to an individual adopting strategy $p$ against an opponent adopting strategy $q$. I now describe how such payoffs can be expressed in matrix notation (Haigh, 1974).

Consider a game with three pure strategies, $H$, $D$ and $R$, and with the following payoff matrix; (payoffs are to the strategy on the left).

$$V = \begin{array}{c} \\ H \\ D \\ R \end{array}\left\{\begin{array}{ccc} H & D & R \\ \alpha_{11} & \alpha_{12} & \alpha_{13} \\ \alpha_{21} & \alpha_{22} & \alpha_{23} \\ \alpha_{31} & \alpha_{32} & \alpha_{33} \end{array}\right.$$

Let strategy $p = p_1\,H + p_2\,D + p_3\,R$, and $q = q_1\,H + q_2\,D + q_3\,R$. Then

$$p'\,Vq = (p_1\,p_2\,p_3)\begin{pmatrix} \alpha_{11} & \alpha_{12} & \alpha_{13} \\ \alpha_{21} & \alpha_{22} & \alpha_{23} \\ \alpha_{31} & \alpha_{32} & \alpha_{33} \end{pmatrix}\begin{pmatrix} q_1 \\ q_2 \\ q_3 \end{pmatrix}$$

$$= (p_1\alpha_{11} + p_2\alpha_{21} + p_3\alpha_{31},\ p_1\alpha_{12} + p_2\alpha_{22} + p_3\alpha_{32},\ p_1\alpha_{13} + p_2\alpha_{23} + p_3\alpha_{33})\begin{pmatrix} q_1 \\ q_2 \\ q_3 \end{pmatrix}$$

$$= \{E(p,H),\ E(p,D),\ E(p,R)\}\begin{pmatrix} q_1 \\ q_2 \\ q_3 \end{pmatrix}$$

$$= q_1\,E(p,H) + q_2\,E(p,D) + q_3\,E(p,R) = E(p,q).$$

Hence $E(p,q)$ can be written $p'\,Vq$.

## B A game with two pure strategies always has an ESS

We can write the payoff matrix:

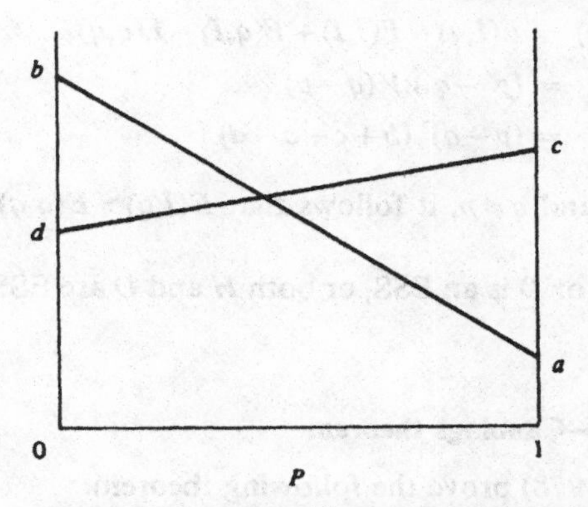


Figure 35. A game with two pure strategies has an ESS.

|   | $H$ | $D$ |
|---|---|---|
| $H$ | $a$ | $b$ |
| $D$ | $c$ | $d$ |

If $a > c$, then $H$ is an ESS.
If $d > b$, then $D$ is an ESS.
If both these inequalities hold, then both $H$ and $D$ are ESS's.

We are left with the case $a < c$ and $d < b$. Let $I$ be the mixed strategy $P(H) + (1-P)(D)$, where $P$ is the probability of playing $H$. If $I$ is an ESS, then by the Bishop–Cannings theorem (Appendix C),

$$aP + b(1-P) = cP + d(1-P).$$

Figure 35 shows that, if $a < c$ and $d < b$, there is always a solution to this equation with $0 < P < 1$. The solution is

$$P = \frac{(b-d)}{(b-d)+(c-a)}.$$

To show that this solution is stable, consider the alternative strategy $q = q(H) + (1-q)(D)$. Since the strategy $I = P(H) + (1-P)(D)$ has the property that $E(H,I) = E(D,I)$, it follows that $E(q,I) = E(I,I)$. Hence $I$ will be stable if $E(I,q) > E(q,q)$.
Now

$$E(I,q)-E(q,q) = E(I,q)-E(I,I)+E(q,I)-E(q,q)$$
$$= (p'-q')\, V(q-p)$$
$$= (p-q)^2\,(b+c-a-d).$$

Since $c>a$ and $b>d$ and $q \neq p$, it follows that $E(I,q) > E(q,q)$, and hence $I$ is stable.

Hence $H$ is an ESS, or $D$ is an ESS, or both $H$ and $D$ are ESS's, or there is a mixed ESS.

## C The Bishop–Cannings theorem

Bishop & Cannings (1978) prove the following theorem.

If $I$ is a mixed ESS with support* $a,b,c\ldots$, then $E(a,I) = E(b,I) = \ldots = E(I,I)$.

*Proof.* Suppose that $a$ is in the support of $I$, and that

$$E(a,I) < E(I,I).$$

Express $I$ as $P(a)+(1-P)(X)$, where $X$ is the strategy, pure or mixed, adopted by $I$ when it is *not a*.

Then

$$E(I,I) = P\,E(a,I)+(1-P)\,E(X,I),$$
$$< P\,E(I,I)+(1-P)\,E(X,I).$$

Hence $$E(I,I) < E(X,I).$$

But this cannot be the case if $I$ is an ESS. Therefore it cannot be the case that $E(a,I) < E(I,I)$. Also since $I$ is an ESS, $E(a,I) \not> E(I,I)$. Hence, for any $a$ in the support of $I$,

$$E(a,I) = E(I,I)$$

and the theorem is proved.

If the strategy set is continuous, and $I$ is a mixed ESS given by the probability density function $p(x)$, then the theorem says that $E(m,I)$ is constant for any value of $m$ for which $p(m) \neq 0$.

This theorem establishes, for evolutionary game theory, the well-known fact that if a mixed strategy, $I$, is the best reply to some

* If pure strategies $a,b,c\ldots$ are played with non-zero probability in the mixed strategy $I$, then $a$, $b$, $c\ldots$ are said to be the 'support' of $I$.

strategy $J$, then $E(a,J)$ is constant for any strategy, $a$, in the support of $I$. It is useful in seeking for candidates for a mixed ESS. Having found such a candidate, it is still necessary to check stability by showing that $E(I,a) > E(a,a)$ for all $a$, where $a$ can be pure or mixed.

## D Dynamics and stability

Suppose that individuals can adopt a number of strategies $i, j \ldots$. Let $p_i, p'_i$ be the frequencies of individuals adopting strategy $i$ in successive generations. If $W_i$ is the fitness of individuals adopting strategy $i$, and $\bar{W}$ is the mean fitness of the population, then

$$\left.\begin{aligned} W_i &= C + \sum_j p_j E(i,j), \\ \bar{W} &= \sum_i p_i W_i, \\ \text{and} \quad p'_i &= p_i W_i / \bar{W}. \end{aligned}\right\} \qquad (D.1)$$

Equations (D.1) describe the dynamics of the population in finite difference form. They can be rewritten:

$$p'_i - p_i = p_i(W_i - \bar{W})/\bar{W}.$$

Provided that the changes per generation are not too great, these can be replaced by the differential equation,

$$dp_i/dt = p_i(W_i - \bar{W})/\bar{W}. \qquad (D.2)$$

Since the right-hand side of the set of equations depicted by (D.2) is divisible by the same function $\bar{W}$, the flows and stationary points are identical for the equations

$$dp_i/dt = p_i(W_i - \bar{W}). \qquad (D.3)$$

It is important to remember that the equivalence of the general equations (D.2) and (D.3) holds only for symmetric games; the differential equation treatment of asymmetric games is considered in Appendix J.

Equations depicted by (D.3) have been suggested by Taylor & Jonker (1978) and Zeeman (1979) to provide continuous dynamics for evolutionary game theory. Identical equations have been used by Eigen & Schuster (1977) to describe concentrations of molecular types during the origin of life. This convergence on the same

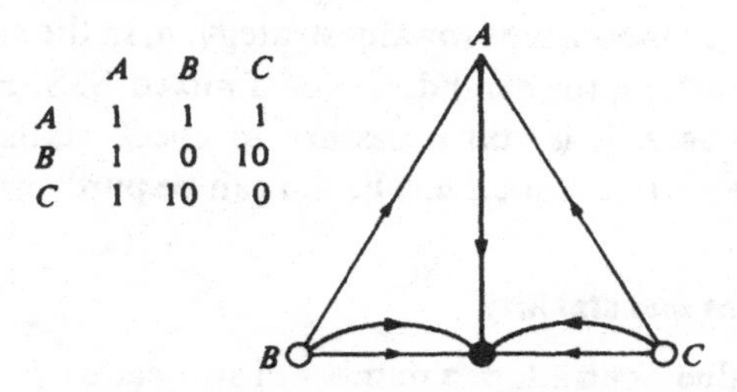

| | *A* | *B* | *C* |
|---|---|---|---|
| *A* | 1 | 1 | 1 |
| *B* | 1 | 0 | 10 |
| *C* | 1 | 10 | 0 |



Figure 36. A game in which one strategy, *A*, is stable against invasion by either *B* alone or *C* alone, but not if *B* and *C* mutants invade simultaneously.

equations is not surprising, since in both cases we are concerned with the evolution of an asexually reproducing population.

Two questions can be asked. First, how far do the conditions (2.4*a*,*b*) for an ESS define the stable states of the dynamical system? Secondly, what differences are there between the behaviour of the finite difference equations (D.1) and the differential equations of type (D.3)?

Considering the first question it is important to realise that equations (D.1) and (D.3) describe the evolution of a population in which only a discrete set of strategies $i, j \ldots$ etc. can exist and breed; a stable state is then a stable genetic polymorphism, or mixture of pure types. In contrast, a mixed strategy satisfying equation (2.4*a*,*b*) can be adopted by a single individual. In what follows, the vector $\tilde{\mathbf{p}}$ will represent the frequencies of types in a polymorphic population, and the vector $\tilde{\mathbf{P}}$ the frequency of actions in a mixed strategy.

The following conclusions hold:

(i) With either type of dynamics, discrete or continuous, if a strategy $\tilde{\mathbf{P}}$ satisfies equation (2.4*a*,*b*) against invasion by any other strategy, pure or mixed, then a population of individuals adopting $\tilde{\mathbf{P}}$ is stable against invasion.

(ii) If only two pure strategies are possible a stable state always exists (Appendix B). If a mixed strategy $\tilde{\mathbf{P}}$ satisfies conditions (2.4*a*,*b*), then both a population of $\tilde{\mathbf{P}}$ individuals and the corresponding polymorphic population $\tilde{\mathbf{p}}$ are stable.

(iii) If more than two pure strategies are possible, and if the dynamics are continuous, then if a mixed strategy $\tilde{\mathbf{P}}$ satisfies equation

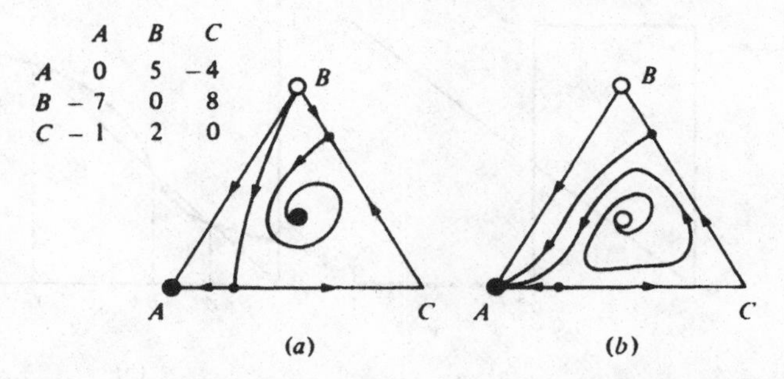


Figure 37. A game for which, if only pure strategies exist, the polymorphism $\tilde{p}=(\frac{1}{3}, \frac{1}{3}, \frac{1}{3})$ is stable, but the corresponding mixed strategy, $\tilde{P}$, can be invaded. (*a*) Dynamics when only pure strategies exist. (*b*) Dynamics when mixed strategies exist; it is supposed that all possible mixed strategies exist in the population, or arise by mutation, and the trajectories show the frequencies with which the acts *A*, *B*, *C* are chosen as the population evolves.

(2.4*a*,*b*) against invasion by any other strategy, pure or mixed, then the corresponding polymorphism $\tilde{p}=\tilde{P}$ will also be stable. This was proved by Taylor & Jonker (1978). and more generally by Zeeman (1979).

Note that it is *not* sufficient for the stability of a polymorphism $\tilde{p}$ that it satisfy conditions (2.4*a*,*b*) only for alternative pure strategies. Thus consider the matrix in Figure 36. Strategy *A* satisfies conditions (2.4*a*,*b*) against invasion by pure *B* or pure *C*. However, as shown in the Figure, *A* is not an attractor of the three-pure-strategy dynamics. Thus *A* cannot be invaded by *B* alone or *C* alone, but can be invaded if *B* and *C* mutants invade simultaneously. This is not a counter-example to the theorem stated above, because *A* does not satisfy conditions (2.4*a*,*b*) against the mixed strategy $(0, \frac{1}{2}, \frac{1}{2})$.

The force of this theorem is that, in a population in which only pure strategies are possible, a population $\tilde{p}$, which may be monomorphic or polymorphic, is stable provided that it satisfies conditions (2.4*a*,*b*) against invasion by any pure or mixed strategy. (I owe my understanding of this point to Dr I. Eshel.)

(iv) The converse to this theorem is not true. Thus a polymorphism, $\tilde{p}$, may be stable although the corresponding mixed strategy, $\tilde{P}$, does not satisfy conditions (2.4*a*,*b*) and can in fact be invaded. As an example, consider the matrix in Figure 37 (Zeeman, 1979). The

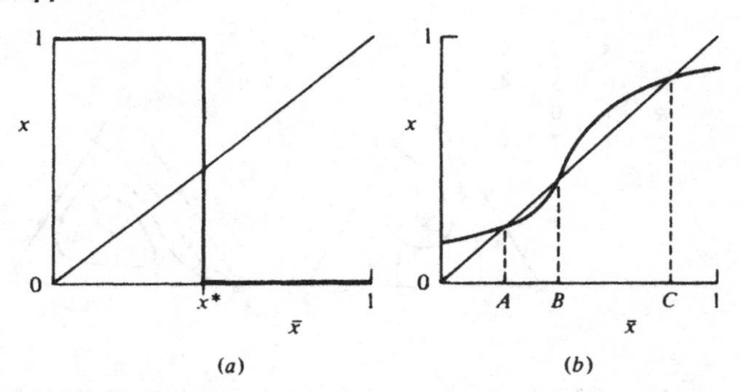


Figure 38. Strong and weak ESS's in the sex ratio game. The bold lines give the optimal sex ratio, $x$, in a population with sex ratio $\bar{x}$. (*a*) Fisher's (1930) original problem; only one ESS, $x^* = 0.5$, exists; (*b*) a case with three ESS's.

dynamics for the three pure strategies only are shown in Figure 37*a*. The state $\tilde{p} = (\frac{1}{3}, \frac{1}{3}, \frac{1}{3})$ is a stable polymorphism, and the pure strategy $A = (1, 0, 0)$ is also an attractor. However, if mixed strategies are allowed, the strategy $\tilde{P} = (\frac{1}{3}, \frac{1}{3}, \frac{1}{3})$ is not stable against invasion by the mixed strategy $(0, \frac{1}{2}, \frac{1}{2})$; if all mixed strategies arise by mutation, it turns out that pure strategy $A$ is the only attractor (Figure 37*b*).

(v) If the dynamics are discrete (equations D.1), and if pure strategies only are possible, then conditions (2.4*a*,*b*) are neither necessary nor sufficient to guarantee the stability of the genetic polymorphism $\tilde{p}$. The matrix in Figure 37 is an example for which the conditions are not necessary, and the Rock–Scissors–Paper game (p. 19) is one for which they are not sufficient.

It will be clear from the above that the relationships between the stability of a polymorphism and of the corresponding mixed strategy are complex. This does not alter the simple conclusion that if a strategy $P$ satisfies conditions (2.4*a*,*b*) for all alternatives, pure or mixed, then $P$ is proof against invasion.

Eshel (1981*a*) and A. Grafen (personal communication) have independently noticed a criterion for the stability of an ESS which is different in kind from those just discussed. Both were seeking evolutionarily stable sex ratios, and the problem they encountered in fact arises only when the strategy set is a continuous one: for example, the interval 0 to 1 in the case of the sex ratio.

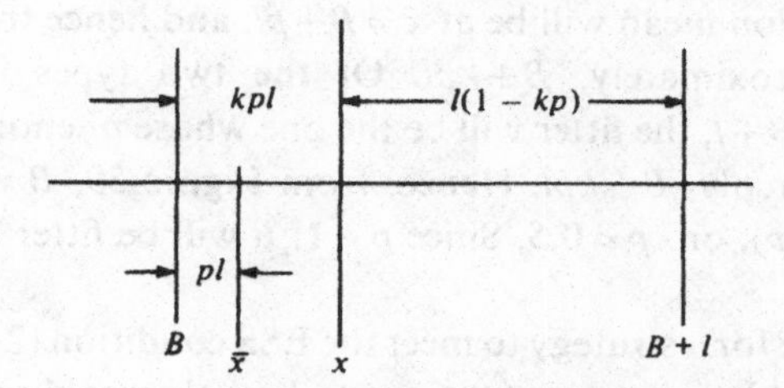


Figure 39. Stability of a weak ESS. For symbols, see text.

One way of formulating the sex ratio problem is as follows. Let $\bar{x}$ be the population sex ratio, and let $x$ be the 'best reply' to $\bar{x}$; that is, $x$ is the sex ratio that maximises the fitness (i.e. expected number of grandchildren) of an individual, given a population ratio of $\bar{x}$. If we plot $x$ against $\bar{x}$ (Figure 38), the ESS value(s), $x^*$, occur where $x = \bar{x}$. Fisher's (1930) original problem is shown in Figure 38*a*; for $\bar{x} < 0.5$ the best reply is $x = 1$ and for $\bar{x} > 0.5$ it is $x = 0$.

The case found by Eshel and Grafen is shown in Figure 38*b*. There are now three ESS's, *A*, *B*, and *C*. The central one, *B*, is only weakly stable, in the following sense. Suppose the population is slightly above *B*; i.e. $\bar{x} = B + \varepsilon$. Then the best reply, $x$, would be $B + k\varepsilon$ where $k = \mathrm{d}x/\mathrm{d}\bar{x}$ at *B*. As drawn, $k > 1$. Hence if the population is $\varepsilon$ away from the ESS at *B*, the best reply is still further from *B*. If a population diverges slightly from *B* (e.g. by drift in a finite population, or because an environmental change alters the payoffs and so moves *B* while leaving $\bar{x}$ unaltered), it will evolve further from *B*, ultimately reaching *A* or *C*, according to the direction of the initial displacement.

The ESS at *B* is 'weakly stable' and those at *A* and *C* are 'strongly stable'; in Eshel's terminology, *A* and *C* are 'continuously stable strategies'. Eshel proves that an ESS is continuously stable if and only if the curve of $x$ against $\bar{x}$ intersects the line $x = \bar{x}$ from above.

Two points are worth adding. First, if the displacement $\varepsilon$ from *B* is small, only mutants of small phenotypic effect (approximately, $< k\varepsilon$) can invade. Secondly, a population at *B* is stable against invasion by any mutant. To see this, consider a population consisting of a proportion $1 - p$ of phenotype *B*, and $p$ mutants of phenotype $B + l$. It is assumed that $p \ll 1$; $l$ can take any value provided $B + l$ is a possible phenotype.

The new population mean will be at $\bar{x} = B + pl$, and hence the new best reply is, approximately, $B + kpl$. Of the two types in the population, $B$ and $B + l$, the fitter will be the one whose phenotype is closest to the best reply, $B + kpl$. Hence, from Figure 39, $B$ will be fitter if $kpl < l(1 - kp)$, or $kp < 0.5$. Since $p \ll 1$, $B$ will be fitter unless $k \gg 1$.

Thus it is possible for a strategy to meet the ESS condition (2.4*a*,*b*), and to be uninvadable by any mutant, yet to be only weakly stable. The possibility arises only if there is a continuous strategy set.

## E Retaliation

The Hawk–Dove game can be extended by the addition of two further pure strategies:

(i) *R*, Retaliator: start by displaying, but escalate if opponent escalates,

(ii) *B*, Bully: start by escalating, but retreat if opponent escalates.

The payoff matrix for the four-strategy game is given in Table 31*a*. This is a simplified version of the game considered by Maynard Smith & Price (1973); in particular, it has been assumed that two Doves (or a Dove and Retaliator, or two Retaliators) can share the resource without the cost of a long contest.

Table 31. *The Hawk–Dove–Retaliator–Bully game*

(*a*)

| | *H* | *D* | *R* | *B* |
|---|---|---|---|---|
| *H* | $v-c$ | $2v$ | $v-c$ | $2v$ |
| *D* | 0 | $v$ | $v$ | 0 |
| *R* | $v-c$ | $v$ | $v$ | $2v$ |
| *B* | 0 | $2v$ | 0 | $v$ |

(*b*)

| | *H* | *D* | *R* | *B* |
|---|---|---|---|---|
| *H* | $v-c$ | $2v$ | $v-c$ | $2v$ |
| *D* | 0 | 0 | 0 | 0 |
| *R* | $v-c$ | 0 | 0 | $2v$ |
| *B* | 0 | $2v$ | 0 | $v$ |

and (*c*) the game as modified by Zeeman (1981)

| | *H* | *D* | *R* | *B* |
|---|---|---|---|---|
| *H* | $v-c$ | $2v$ | $v-c+\varepsilon$ | $2v$ |
| *D* | 0 | $v$ | $v-\varepsilon$ | 0 |
| *R* | $v-c-\varepsilon$ | $v+\varepsilon$ | $v$ | $2v$ |
| *B* | 0 | $2v$ | 0 | $v$ |

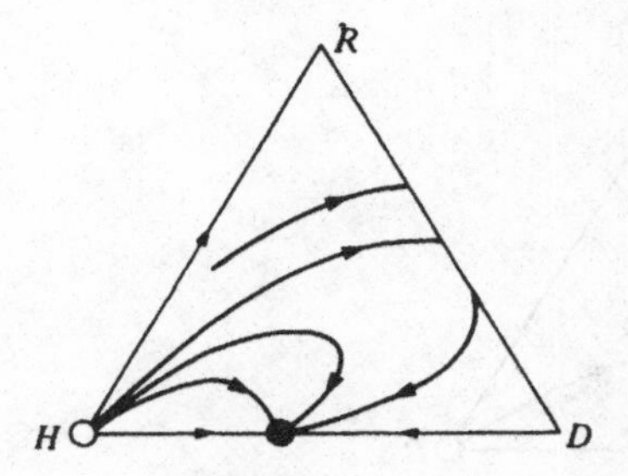


Figure 40. The Hawk–Dove–Retaliator game.

Several difficulties and errors have arisen in the analysis of this game:

(i) Gale and Eaves (1975) pointed out that Maynard Smith & Price had missed an ESS of their game, consisting of a mixture of *H* and *B*.

(ii) In the absence of *H* and *B*, there is no difference between *R* and *D*. This selective neutrality introduces difficulties into the analysis.

(iii) *R* is not an ESS of the three-strategy *H–D–R* game.

(iv) P. G. Caryl (personal communication) has pointed out a biological implausibility of the original game. This is the assumption that two Doves can settle a contest at a small cost because of time-wasting. If the resource is divisible, this could be true, so that Table 31*a* can be thought of as referring to a contest for a divisible resource. If the resource is indivisible, however, a contest between two Doves becomes a war of attrition, in which the expected cost is equal to $V/2 = v$. If so, the payoff matrix is as shown in Table 31*b*. The criticism of the original game is valid, but it turns out that the change from *a* to *b* in Table 31 does not alter the dynamics.

I first analyse the three-strategy *H–D–R* game, and then turn to the full four-strategy game. The dynamics for the *H–D–R* game are shown in Figure 40. A polymorphism with pure *H* and pure *D* is the only ESS.

There is, however, a good reason for not accepting this as a complete account of the game. If the fourth strategy, *B*, is present, it is no longer true that *R* is always eliminated. This is because *R* does better against *B* than *D* does. Unfortunately, if we introduce the fourth strategy, *B*, then no stable state exists. Instead, the system cycles indefinitely *R*→*RD* line→*HD* polymorphism→*HB* polymor-

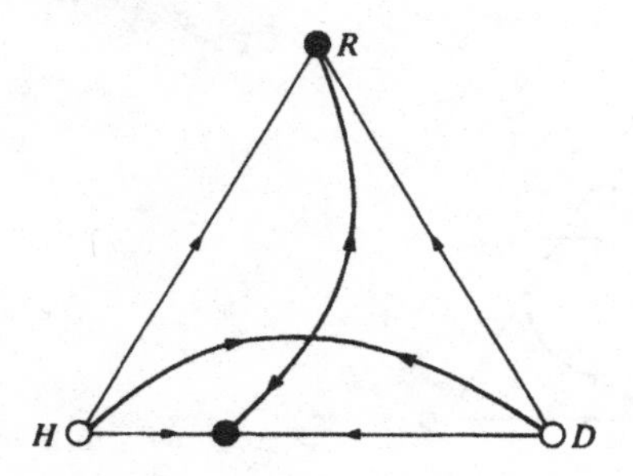


Figure 41. The modified Hawk–Dove–Retaliator game.

phism → $R$. This anomalous behaviour arises from the rather implausible assumption that $R$ and $D$ are identical when $H$ and $B$ are absent.

Zeeman (1981) suggests that the matrix should be modified to allow for the fact that, in an $R$ v. $D$ contest, the Retaliator will sometimes discover that its opponent will never escalate, and will exploit the discovery. He also allows for a slight advantage of $H$ against $R$, on the grounds that $H$ has the advantage of escalating first. Both these effects were in fact present in the original Maynard Smith & Price model. The resulting payoff matrix is given in Table 31*c* (an equivalent modification could be made to Table 31*b*, but the dynamics would be the same).

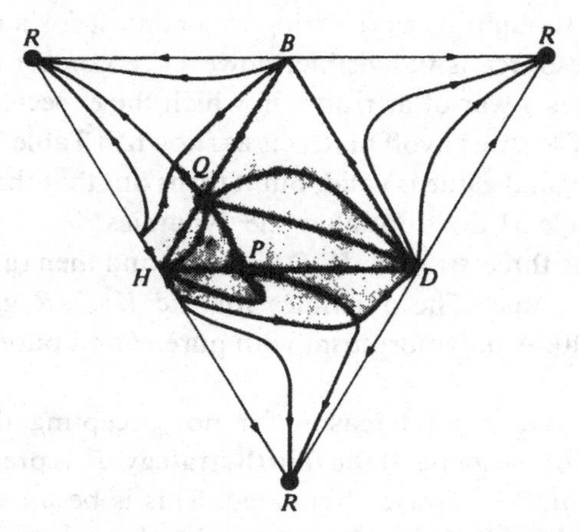


Figure 42. The Hawk–Dove–Retaliator–Bully game. The diagram should be imagined folded into a tetrahedron along the lines *HB*, *BD*, *DH*. (After Zeeman, 1981).

The dynamics of the modified *H–D–R* game are shown in Figure 41. There are two alternative ESS's, Retaliator and a Hawk–Dove polymorphism. The dynamics of the full four-strategy game are shown in Figure 42. There are two ESS's, Retaliator and a Hawk–Bully polymorphism; this confirms the results obtained by Gale & Eaves (1975) by simulating the original matrix given by Maynard Smith & Price.

It is somewhat ironic that I first proposed the idea of the ESS to formalise a verbal argument by George Price that the prevalence of ritualised behaviour in animal contests could be explained by the phenomenon of retaliation (a similar proposal had been made by Geist, 1966). Although I think we were essentially correct, I certainly got many of the details wrong.

## F Games between relatives

If a trait affects the survival or fertility of relatives of the individual possessing it, its evolution can be modelled in one of two ways (Hamilton, 1964; Orlove, 1979; Cavalli-Sforza & Feldman, 1978); briefly, the 'inclusive fitness' of an individual, ego, allows for the assistance which ego gives to his relatives, whereas the 'neighbour-modulated' or 'personal' fitness counts only ego's direct offspring, but allows for the help ego receives from his relatives. These two approaches have been applied to games between relatives. The personal fitness approach is formally correct, but does not provide a simple way of finding ESS's, whereas the inclusive fitness method does provide a means of finding ESS's but can lead to wrong conclusions.

The model is as follows. Fitness is determined by a series of pairwise contests, and reproduction is asexual, with like begetting like, exactly as in the basic model (Chap. 2). Instead of pairs being randomly formed, however, a fraction $r$ of all contests are between members of the same clone (and hence between opponents with the same pure or mixed strategy), and a fraction $(1-r)$ are between randomly assorted opponents. Thus $r$ is the probability that two opponents will be identical by descent, and so has some resemblance to the coefficients of relationship in a sexual population.

Grafen (1979) has taken the personal fitness approach, as follows.

Table 32. *The Hawk–Dove game between relatives*

| (a) Basic matrix | | | | | (b) Derived inclusive fitness matrix | | |
|---|---|---|---|---|---|---|---|
| | Player 2 | | | | | | |
| | | $H$ | $D$ | | | $H$ | $D$ |
| Player 1 | $H$ | $a$ | $b$ | $\xrightarrow{r}$ | $H$ | $a(1+r)$ | $b+rc$ |
| | $D$ | $c$ | $d$ | | $D$ | $c+rb$ | $d(1+r)$ |

Let $W(x)$ be the personal fitness of an $x$-strategist in a population whose average strategy is $\alpha$. Then

$$W(x) = W_0 + r\,E(x,x) + (1-r)\,E(x,\alpha).$$

Hence, if $I$ is an ESS and $m$ a rare mutant,

$$W(I) = W_0 + rE(I,I) + (1-r)\,E(I,I) = W_0 + E(I,I)$$

and

$$W(m) = W_0 + rE(m,m) + (1-r)\,E(m,I).$$

Hence, since $I$ is an ESS, $W(I) \geqslant W(m)$ for all $m \neq I$, or

$$E(I,I) \geqslant rE(m,m) + (1-r)E(m,I). \qquad \text{(F.1)}$$

(F.1) is a necessary condition for $I$ to be an ESS; but the condition does not help to find candidate strategies. This can best be done by the inclusive fitness approach. For example, consider the two-strategy game shown in Table 32. From this, it is possible to write down a derived matrix, whose entries are the changes in inclusive fitness of player 1. It is then easy, by applying the Bishop–Cannings theorem (Appendix C) to find the ESS of the derived matrix. In this case,

if $a(1+r) > c+br$, $H$ is an ESS,

if $d(1+r) > b+cr$, $D$ is an ESS,

and if neither inequality holds, there is a mixed ESS with $PH$ and $(1-P)D$, where

$$P = \frac{(b-d)+r(c-d)}{(1+r)\,(b+c-a-d)}. \qquad \text{(F.2)}$$

Table 33. *The derived inclusive fitness matrix for a two-strategy game*

Case 1

| | *H* | *D* | | | *H* | *D* |
|---|---|---|---|---|---|---|
| *H* | 1 | 6 | $\xrightarrow{r=\frac{1}{2}}$ | *H* | 1.5 | 7 |
| *D* | 2 | 3 | | *D* | 5 | 4.5 |

The ESS is $P(H)=0.75$ between non-relatives, and $P(H)=0.417$ between relatives ($r=\frac{1}{2}$).

Case 2

| | *H* | *D* | | | *H* | *D* |
|---|---|---|---|---|---|---|
| *H* | 7 | 8 | $\xrightarrow{r=\frac{1}{2}}$ | *H* | 10.5 | 11 |
| *D* | 6 | 9 | | *D* | 10 | 13.5 |

Both *H* and *D* are ESS's of the inclusive fitness matrix, but *H* does not satisfy conditions (F.1), because it can be invaded by *D*; $rE(m,m)+(1-r)E(m,I)$ $=\frac{1}{2}\times 9+\frac{1}{2}\times 6=7.5>E(H,H)$.

Hines & Maynard Smith (1979) have shown that any strategy which is an ESS by Grafen's condition (F.1) is also an ESS of the inclusive fitness matrix, but the reverse is not true. The recommended procedure, therefore, is to seek candidates for the ESS by applying the Bishop–Cannings theorem of equal payoffs to changes in *inclusive* fitness, and to check that the resulting strategies satisfy equation (F.1).

Table 33 shows two numerical examples. Case 1 confirms the common-sense expectation that the ESS in games between relatives is more cooperative than in games between non-relatives. Case 2 is an example in which an ESS of the inclusive fitness matrix is not stable by condition (F.1).

In games between non-relatives, if the vector $\tilde{\boldsymbol{P}}$ is a mixed ESS, then, if only pure strategists can exist, the vector $\tilde{\boldsymbol{p}}$ represents a stationary polymorphism (which is necessarily stable if there are only two pure strategies). This does not hold for games between relatives. Thus for the game in Table 32*b*, if only pure *H* and pure *D* exist, the frequency of *H* at equilibrium is

$$p = \frac{b-d+r(a-b)}{(1-r)(b+c-a-d)}, \tag{F.3}$$

which is not the same as (F.2).

It would be desirable to extend the treatment of games between relatives to a sexual diploid population. Unfortunately, it turns out that, when opponents are relatives, a treatment similar to that on p. 40 is difficult.

## G The war of attrition with random rewards

A population contains two types of individual; type 1, frequency $q$, receives payoff $v_1$ for winning, and type 2, frequency $1-q$, receives payoff $v_2$ for winning. Each individual knows its own type, and the values of $v_1$, $v_2$ and $q$, but not the actual type of its opponent. For example, if animals are hungry (type 1) or not (type 2), and $v_1$ and $v_2$ are the values of a food item to a hungry and well-fed animal, respectively, then an animal is assumed to know whether it is hungry, but only the probability that its opponent is hungry.

Table 34. *Payoffs for the war of attrition with random rewards*

| | To type 1 | To type 2 |
|---|---|---|
| $m_1 > m_2$ | $v_1-m_2$ | $-m_2$ |
| $m_1 = m_2$ | $v_1/2-m_1$ | $v_2/2-m_2$ |
| $m_1 < m_2$ | $-m_1$ | $v_2-m_1$ |

Contests take place between randomly chosen individuals. In a contest between an individual of type 1 who chooses a permissible cost $m_1$ and an individual of type 2 who chooses $m_2$, the payoffs are given in Table 34. Let the ESS be to choose $x$ with probability density $p_1(x)$ if type 1, and $p_2(x)$ if type 2. A randomly selected opponent then plays $x$ with probability density

$$G(x) = q\,p_1(x)+(1-q)p_2(x). \tag{G.1}$$

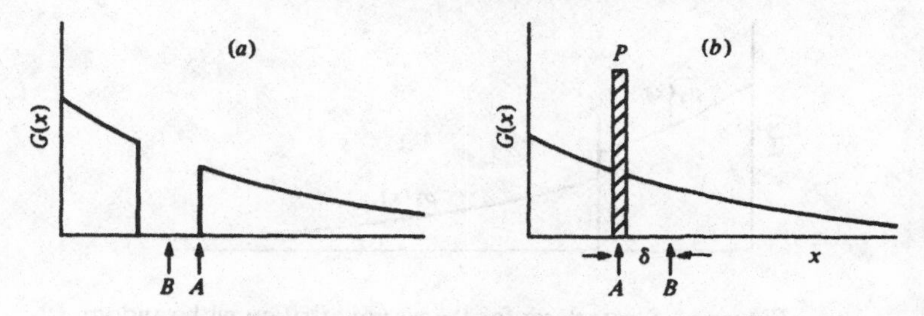


**Figure 43. Proof that there can be no gaps and no atoms of probability in the strategy set for the war of attrition with random rewards.**

It will first be shown that $G(x)$ has no 'gaps' and no 'atoms of probability'. Suppose that $G(x)$ has a gap, as in Figure 43*a*. Compare the payoffs to an individual choosing $A$ and choosing $B$. The two choices will win on the same occasions, and on those occasions will receive the same payoff. They also lose on the same occasions, and then choice $A$ pays out more than $B$. Thus choice $B$ does better than choice $A$. Hence $G(x)$ cannot be an ESS. That is, an ESS cannot have a gap.

Now consider Figure 43*b*, in which there is a non-zero probability, $P$, of choosing $A$. Compare the payoffs to individuals choosing $A$, and $B = A + \delta$. On those occasions when both choices win, their payoffs are the same. When both lose, $B$'s payoff is less by a quantity $\delta$. However, $B$ wins a fraction $P/2$ of contests which are lost by $A$ (i.e. half the contests against $A$ opponents). Since $P$ is non-zero, it follows that, if $\delta$ is sufficiently small, the expected payoff to $B$ is greater than to $A$. But if $G(x)$ is an ESS, this is ruled out by the Bishop–Cannings theorem (Appendix C). Hence $G(x)$ cannot have an atom of probability.

It will now be shown that $p_1(x)$ and $p_2(x)$ cannot overlap. Consider the strategy $J = [m, p_2(x)]$; that is, choose a fixed $m$ if type 1, choose $p_2(x)$ if type 2. $I$ is the ESS, $[p_1(x), p_2(x)]$. Then

$$E(J,I) = q\,[v_1 P(m) - R(m)] + (1-q)\,S,$$

where $P(m)$ is the probability that $m$ wins against $G(x)$, $R(m)$ is the expected cost of choosing $m$ against $G(x)$, and $S$ is the expected payoff of type 2 against $G(x)$.

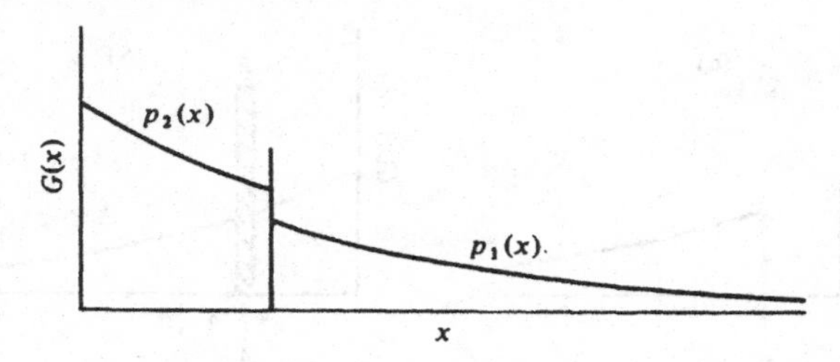


Figure 44. Strategy set for the war of attrition with random rewards. .

By the Bishop–Cannings theorem, $E(J,I)$ is a constant for all $m$ in the support of $p_1(x)$. Hence

$$v_1P(m) - R(m) = A,$$

where $A$ is a constant and $P(m)$ and $R(m)$ are functions of $m$ which do not depend on whether the contestant is type 1 or type 2.

By an exactly similar argument, if $m$ is in the support of $p_2(x)$,

$$v_2P(m) - R(m) = B,$$

where $B$ is a constant. Hence, if $m$ is in the support of both $p_1(x)$ and $p_2(x)$,

$$(v_1 - v_2)\ P(m) = A - B. \tag{G.2}$$

Since $G(x)$ has no gaps, $P(m)$ increases monotonically with $m$. Since $v_1 \neq v_2$, it follows that only one value of $m$ can satisfy (G.2). That is, there can be no overlap between $p_1(x)$ and $p_2(x)$. Since there are no gaps, the two distributions meet at $m$.

It can further be shown that, if $v_1 > v_2$, then $p_1(x)$ falls above $p_2(x)$, as in Figure 44. These conclusions can be extended to the case with more than two types, with different payoffs for winning. In the limit, if the payoff for winning is continuously distributed, there will be a unique choice associated with each payoff, victory always going to the contestant with the higher payoff.

Bishop, Cannings & Maynard Smith (1978) gave more formal proofs of these assertions, and show how the actual probability distributions can be derived. It is further shown that $G(x)$ is strictly decreasing in contests of this type.

## H The ESS when the strategy set is defined by one or more continuous variables

In the sex ratio problem (p. 43), the possible values of the sex ratio, $s$, are assumed to vary continuously from 0 to 1. In the anisogamy problem (p. 47), the possible mass of a gamete is assumed to vary continuously from $\delta$ to some maximum value $M$. In each case, we seek a unique evolutionarily stable value, $s^*$ and $m^*$ respectively, such that a population of individuals with this value cannot be invaded by mutants with any different value.

To find $m^*$, we must first find the fitness, $W(m,m^*)$, of a rare individual adopting strategy $m$ in a population adopting strategy $m^*$. The method of finding $W(m,m^*)$ will depend on the particular problem. For completeness we should write the fitness as $W(m,m^*)$ rather than as $W(m)$ because, if fitnesses are frequency-dependent, $W$ will depend on both $m$ and $m^*$; in practice, however, the shorter notation can be used.

If $m^*$ is to be evolutionarily stable then $W(m^*,m^*) > W(m,m^*)$ for all $m \neq m^*$. If $W$ is a differentiable function, this is satisfied if

$$\left.\begin{aligned} \left[\frac{\partial W(m,m^*)}{\partial m}\right]_{m=m^*} &= 0, \\ \left[\frac{\partial^2 W(m,m^*)}{\partial m^2}\right]_{m=m^*} &< 0. \end{aligned}\right\} \qquad \text{(H.1)}$$

These conditions guarantee stability only against mutants of small phenotypic effect; that is, $m \simeq m^*$. When an $m^*$ satisfying (H.1) has been found, it is then necessary to check that it is stable against more extreme mutants (see, for example, p. 52).

The method can readily be extended to cases where a strategy set requires more than one continuous variable for its description. Thus suppose that the strategy is defined by two variables, $x$ and $y$, and that $x^*$, $y^*$ is an ESS. We consider the stability of this ESS against invasion by mutants affecting only one of the variables at a time. Thus let $W(x\ y^*, x^*\ y^*)$ be the fitness of an $xy^*$ mutant in an $x^*y^*$ population. Stability requires that

$$[\partial W(xy^*, x^*y^*)/\partial x]_{x=x^*} = 0$$

and $$[\partial^2 W(xy^*, x^*y^*)/\partial x^2]_{x=x^*} < 0,$$

and a similar requirement on $x^*y$ mutants.
The two stationarity conditions give two equations which can be solved for $x^*$ and $y^*$.

As an example (Maynard Smith, 1980), the method was used to find the evolutionarily stable parental investment in sons ($m^*$) and daughters ($f^*$), subject to the constraints of a 1 : 1 sex ratio and a total investment $n(m^*+f^*)=\text{const.}$, where $2n$ is the family size.

## I To find the ESS from a set of recurrence equations

On p. 45 we obtained recurrence equations of the form

$$\left.\begin{aligned} p_{n+1} &= \alpha p_n + \beta P_n, \\ P_{n+1} &= \gamma p_n + \delta P_n. \end{aligned}\right\} \qquad (I.1)$$

These equations have the solution

$$P_n = A\lambda_1^n + B\lambda_2^n,$$

where $A$ and $B$ are constants, and $\lambda_1$, $\lambda_2$ are the solutions of the characteristic equation,

$$\lambda^2 - (\alpha+\delta)\lambda + (\alpha\delta - \beta\gamma) = 0. \qquad (I.2)$$

The parameters $\alpha$, $\beta$, $\gamma$, $\delta$ are functions of a variable $a$, defining the phenotype of a mutant, and of $a^*$, defining the phenotype at the ESS. (Since we were concerned only with phenotypes on the boundary of the set, we need not include the variable $b$, since $b=f(a)$ and $b^*=f(a^*)$.) Thus (I.2) can be written

$$\phi(\lambda, a, a^*) = 0. \qquad (I.3)$$

The eigenvalue $\lambda$ measures the rate of increase of the mutant (phenotype $a$) relative to the typical (phenotype $a^*$). It follows that:

(i) when $a = a^*$, $\lambda = 1$, and
(ii) since $a^*$ is an ESS, $\lambda < 1$ for $a \neq a^*$.

Hence we seek a value of $a^*$ such that

$$\left(\frac{d\lambda}{da}\right)_{a=a^*} = 0; \left(\frac{d^2\lambda}{da^2}\right)_{a=a^*} < 0. \quad (I.4)$$

From the constraint equation (I.3) we have

$$d\phi = \frac{\partial\phi}{\partial\lambda} d\lambda + \frac{\partial\phi}{\partial a} da = 0,$$

or
$$\frac{d\lambda}{da} = -\frac{\partial\phi}{\partial a}\bigg/\frac{\partial\phi}{\partial\lambda}.$$

Hence, if $\partial\phi/\partial a = 0$, then $d\lambda/da = 0$.

Therefore, the method of finding $a^*$ is to solve the equation

$$\left(\frac{\partial\phi}{\partial a}\right)_{\substack{a=a^* \\ \lambda=1}} = 0. \quad (I.5)$$

For example, the recurrence equations on p. 45 have the characteristic equation

$$\lambda^2 - \tfrac{1}{2}\left(1+\frac{b}{b^*}\right)\lambda + \tfrac{1}{4}\left(\frac{b}{b^*} - \frac{a}{a^*}\right) = 0.$$

Hence
$$\frac{\partial\phi}{\partial a} = -\frac{1}{2b^*}\frac{db}{da}\lambda + \frac{1}{4b^*}\frac{db}{da} - \frac{1}{4a^*},$$

and putting $\lambda = 1$, (I.5) gives

$$\frac{1}{a^*} + \frac{1}{b^*}\left(\frac{db}{da}\right)_* = 0,$$

which is the result we got on p. 45 by a less general method.

Thus to find the ESS, given a set of recurrence equations, it is sufficient to write down the characteristic equation, and then to solve equation (I.5).

## J Asymmetric games with cyclic dynamics

Consider an asymmetric game in which two pure strategies are possible in each role. The payoff matrix is given in Table 35. Suppose

Table 35. *Payoff matrix for the two-pure-strategy asymmetric game*

| | | Role 2 | |
|---|---|---|---|
| | | $R$ | $S$ |
| Role 1 | $A$ | $a$ \ $r$ | $c$ \ $s$ |
| | $B$ | $b$ \ $t$ | $d$ \ $u$ |

that only pure strategies exist. Let their frequencies be $P(A) = X$ and $P(R) = Y$. Then the fitnesses of the pure strategies are:

$$W(A) = aY + c(1-Y), \qquad V(R) = rX + t(1-X).$$
$$W(B) = bY + d(1-Y), \qquad V(S) = sX + u(1-X).$$

The mean fitnesses are:

$$\bar{W} = XW(A) + (1-X)W(B); \qquad \bar{V} = YV(R) + (1-Y)V(S).$$

Then proceeding as in Appendix D, the differential equations for $X$ and $Y$ are

$$\frac{\mathrm{d}X}{\mathrm{d}t} = X\left(\frac{W(A)-\bar{W}}{\bar{W}}\right); \ \frac{\mathrm{d}Y}{\mathrm{d}t} = Y\left(\frac{V(R)-\bar{V}}{\bar{V}}\right). \qquad \text{(J.1)}$$

Schuster & Sigmund (1981), basing themselves on Dawkins' (1976) 'battle of the sexes' described on p. 130, have considered this problem. Instead of equations (J.1), however, they used the equations

$$\mathrm{d}X/\mathrm{d}t = X[W(A)-\bar{W}]; \ \mathrm{d}Y/\mathrm{d}t = Y[V(R)-\bar{V}]. \qquad \text{(J.2)}$$

It was argued in Appendix D that, in symmetric games, it is legitimate to replace equations of the form (J.1) by (J.2), because the stationary points and flows will remain unaltered. This is no longer true, however, for asymmetric games, because $\bar{V} \neq \bar{W}$. There is

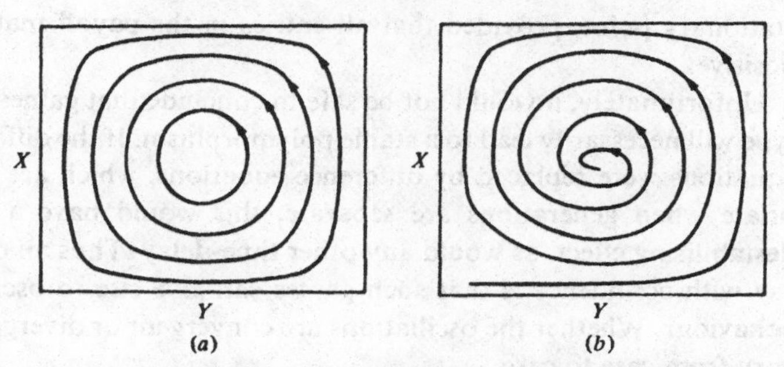


Figure 45. Dynamics for the two-strategy asymmetric game; (*a*) for equations (J.2); (*b*) for equations (J.1).

therefore room for doubt as to which form is more appropriate; they will be considered in turn.

The interesting case, with an internal stationary point and cyclical behaviour, arises when $r > s$, $b > a$, $u > t$, and $c > d$ (or when all these inequalities are reversed). Qualitatively, nothing is lost if we consider a simple numerical example, with $r = b = u = c = 2$ and $s = a = t = d = 1$. By symmetry, the stationary point is at $X = 0.5$, $Y = 0.5$. Writing $X = 0.5 + x$; $Y = 0.5 + y$, equations (J.2) become

$$\frac{dx}{dt} = -\frac{y}{3}(1-4x^2); \quad \frac{dy}{dt} = \frac{x}{3}(1-4y^2). \tag{J.3}$$

If we ignore the terms in $x^2$ and $y^2$, equations (J.3) describe simple harmonic motion. The complete equations also describe a conservative system, because $H = x^2 + y^2 - 4x^2y^2$ is a constant of the motion. Thus (J.3) describe a series of closed loops (Figure 45*a*).

Biologists have been taught to be distrustful of conservative systems. It is therefore some comfort that equations (J.1) are asymptotically stable. For the numerical example, the equations become

$$\frac{dx}{dt} = -\frac{y}{3}\frac{(1-4x^2)}{(1-4xy/3)}; \quad \frac{dy}{dt} = \frac{x}{3}\frac{(1-4y^2)}{(1+4xy/3)}. \tag{J.4}$$

Joseph Hofbauer has shown me that, for these equations, $H = x^2 + y^2 - 4x^2y^2$ is a Lyapunov function. That is, $dH/dt \leqslant 0$. Hence the internal stationary point is asymptotically stable (Figure 45*b*). Hofbauer has further shown that equations (J.1) converge to the stationary point, provided that all entries in the payoff matrix are positive.

Unfortunately, it would not be safe to conclude that games of this type will necessarily lead to a stable polymorphism. If the differential equations were replaced by difference equations, which are appropriate when generations are separate, this would have a strong destabilising effect, as would any other time-delay. Thus all one can say with confidence is that such games can give rise to oscillatory behaviour. Whether the oscillations are convergent or divergent will vary from case to case.

### K The reiterated Prisoner's Dilemma

The Prisoner's Dilemma game is shown in Table 36. A match consists of a sequence of games between the same two players. After each game, there is a probability $w$ of a further game. Thus the expected number of games per match is $1+w+w^2+\ldots=1/(1-w)$

Table 36. *The Prisoner's Dilemma game*

| | | Player $B$ | |
|---|---|---|---|
| | | Cooperate $C$ | Defect $D$ |
| Player $A$ | Cooperate $C$ | $R$ | $S$ |
| | Defect $D$ | $T$ | $P$ |

The game is defined by $T>R>P>S$. It is also assumed that $2R>T+S$; this ensures that the payoff is greater to each of two players who cooperate than to a pair who alternately cooperate and defect.

The strategy TIT FOR TAT, or *TFT*, plays $C$ on the first game, and subsequently plays whatever its opponent played on the previous game. Axelrod (1981) shows that *TFT* is an ESS against any alternative, provided that enough games are played.

We first show that the only alternatives we need to consider are *CCCC* . . . ; *DDDD* . . . ; and *DCDC* . . . , because no other strategy can do better against *TFT* than all of these.

Note first that *TFT* has a memory of only one game, and that the expected number of further games at any time is constant. If $I$ is a strategy played against $TFT$, a play of $C$ by $I$ at any point will reset the match to exactly the state it was in initially. Also, if the first play by $I$ is $D$, a play of $D$ at any later point will reset the match to its initial state.

Note next that if $I$ is a best reply to $TFT$, then it must make the same play as it did on the first move if the initial state ever recurs; if there were a better play in that state, it should have been played initially.

It then follows that a best reply must have one of three forms:

(i) First play $C$; the initial state is then repeated on the second game, and $C$ must be played again, and so on. That is, play $CCCC \ldots$.

(ii) First play $D$, and then $C$; the initial state is then repeated on the third game, and $D$ must be played again, and so on. That is, play $DCDCDC \ldots$.

(iii) First play $D$, and then $D$; the initial state is repeated on the third game, and $D$ must be played. That is play $DDDD \ldots$.

It is now easy to see in what circumstances one of these alternatives can invade.

The payoff to $TFT$ against itself is $R + wR + w^2R + \ldots = R/(1-w)$.

The payoff to $CCCC \ldots$ against $TFT$ is the same, so $CCCC \ldots$ cannot invade.

The payoff to $DCDCDC \ldots$ against $TFT$ is $T + wS + w^2T + w^3S + \ldots = (T + wS)/(1-w^2)$.

The payoff to $DDDD \ldots$ against $TFT$ is $T + wP + w^2P + \ldots = T + wP/(1-w)$.

Hence $TFT$ is an ESS provided that $R/(1-w) \geqslant (T+wS)/(1-w^2)$ and $R/(1-w) \geqslant T + wP/(1-w)$. This reduces to

$$w \geqslant \frac{T-R}{R-S} \text{ and } w \geqslant \frac{T-R}{T-P}.$$

Hence TIT FOR TAT is an ESS provided $w$ is large enough.

# Explanation of main terms

A 'strategy' is a specification of what an individual will do in any situation in which it may find itself. A strategy may be 'pure' or 'mixed'; in the latter case there is a random element in the specification.

An 'action' is a behaviour performed in a particular situation.

An 'asymmetric' contest is one in which there is a difference in 'role' between the contestants, of a kind which enables either of them to adopt a strategy 'in role 1, do *A*; in role 2, do *B*'. Clearly, roles must be perceived by the contestants; otherwise, they could not affect behaviour. Examples are male and female, or owner and intruder. It is assumed that the circumstances, genetic or environmental, which decide in which role an individual finds itself, act independently of genes determining its strategy.

A 'symmetric' contest is one with no role differentiation.

A 'payoff', written $E(A,B)$, is the expected change of fitness of an individual adopting a strategy *A* against an opponent adopting *B*.

A population is said to be in an 'evolutionarily stable state' if its genetic composition is restored by selection after a disturbance, provided the disturbance is not too large. Such a population can be genetically monomorphic or polymorphic.

An 'ESS' or 'evolutionarily stable strategy' is a strategy such that, if all the members of a population adopt it, no mutant strategy can invade. For the extended model of games against the field, an ESS must satisfy equations (2.9). For pairwise contests in an infinite asexual population, an ESS must satisfy conditions (2.4*a*,*b*). Two points about this definition should be noted. First, if stability requires a mixture of pure strategies, then individuals must adopt the appropriate mixed strategy; a genetically polymorphic population may be in an evolutionarily stable state, but, strictly, no individual is adopting an ESS. Secondly, I have preferred to define an ESS as an uninvadable strategy, rather than as a strategy satisfying any particular mathematical conditions. In some cases, however, it is convenient to use the term ESS for any strategy satisfying conditions (2.4*a*,*b*); I hope that the context will make it clear in which sense the term is being used.

# References

Abegglen, J.-J. (1976). On socialization in hamadryas baboons. Ph.D. thesis, University of Zurich.

Alcock, J., Jones, C.E. & Buckman, S.L. (1977). Male nesting strategies in the bee *Centris pallida* Fox (Anthophoridae: Hymenoptera). *Am. Nat.* **111,** 145–55.

Andersson, M. (1980). Why are there so many threat displays? *J. theor. Biol.* **86,** 773–81.

Axelrod, R. (1981). The emergence of cooperation among egoists. *Am. political Sci. Rev.* (in press).

Axelrod, R. & Hamilton, W.D. (1981). The evolution of cooperation. *Science*, **211,** 1390–6.

Bachmann, C. & Kummer, H. (1980). Male assessment of female choice in hamadryas baboons. *Behav. ecol. Sociobiol.* **6,** 315–21.

Baker, M.C. (1978). Flocking and feeding in the great tit *Parus major* – an important consideration. *Am. Nat.* **112,** 779–81.

Baker, R.R. (1972). Territorial behaviour of the nymphalid butterflies, *Aglais urticae* (L.) and *Inachis io* (L.). *J. Anim. Ecol.* **41,** 453–69.

Baldwin, B.A. & Meese, G.B. (1979). Social behaviour in pigs studied by means of operant conditioning. *Anim. Behav.* **27,** 947–57.

Balph, M.H. & Balph, D.F. (1979). On the relationship between plumage variability and social behaviour in wintering pine siskins (*Carduelis pinus*). Paper presented at XVIth International Ethological Conference.

Balph, M.H., Balph, D.F. & Romesburg, H.C. (1979). Social status signalling in winter flocking birds: an examination of a current hypothesis. *The Auk*, **96,** 78–93.

Barnard, C.J. & Sibly, R.M. (1981). Producers and scroungers: a general model and its application to captive flocks of house sparrows. *Anim. Behav.* **29,** 543–50.

Bateman, A.J. (1948). Intra-sexual selection in *Drosophila. Heredity*, **2,** 349–68.

Bertram, B.C.R. (1976). Kin selection in lions and in evolution. In *Growing Points in Ethology*, ed. P.P.G. Bateson & R.A. Hinde, pp. 281–301. Cambridge University Press.

Birky, C.W. (1978). Transmission genetics of mitochondria and chloroplasts. *A. Rev. Genet.* **12,** 471–512.

Bishop, D.T. & Cannings, C. (1978). A generalised war of attrition. *J. theor. Biol.* **70,** 85–124.

Bishop, D.T., Cannings, C. & Maynard Smith, J. (1978). The war of attrition with random rewards. *J. theor. Biol.* **74,** 377–88.

Brockmann, H.J. & Dawkins, R. (1979). Joint nesting in a digger wasp as an evolutionarily stable preadaptation to social life. *Behaviour*, **71,** 203–45.

Brockmann, H.J., Grafen, A. & Dawkins, R. (1979). Evolutionarily stable nesting strategy in a digger wasp. *J. theor. Biol.* **77,** 473–96.

Bruning, D.F. (1973). The greater rhea chick and egg delivery route. *Nat. Hist.*, **82,** 68–75.

Bull, J.J. (1980). Sex determination in reptiles. *Q. Rev. Biol.* **55,** 3–21.

Burgess, J.W. (1976). Social spiders. *Scient. Am.* March, 100–6.

Bush, R.R. & Wilson, T.R. (1956). Two-choice behaviour of a paradise fish. *J. exp. psychol.* **51,** 315–22.

Cade, W. (1979). The evolution of alternative male reproductive strategies in field crickets. In *Sexual Selection and Reproductive Competition in Insects*, ed. M.S. & N.A. Blum, pp. 343–79. Academic Press: New York.

Caryl, P.G. (1979). Communication by agonistic displays: what can game theory contribute to ethology? *Behaviour*, **68,** 136–69.

Cavalli-Sforza, L.L. & Feldman, M.W. (1978). Darwinian selection and 'altruism'. *Theor. Pop. Biol.* **14,** 268–80.

Charlesworth, B. (1977). Population genetics, demography and the sex ratio. In *Measuring Selection in Natural Populations*, ed. F.B. Christiansen & T.M. Fenchel, pp. 345–63. Springer-Verlag: Berlin.

– (1980). *Evolution in Age-structured Populations*. Cambridge University Press.

Charnov, E.L. (1979). The genetical evolution of patterns of sexuality. *Am. Nat.* **113,** 465–80.

– (1981). *The Theory of Sex Allocation*. Princeton University Press (in press).

Charnov, E.L. & Bull, J.J. (1977). When is sex environmentally determined? *Nature, Lond.* **266,** 828–30.

Charnov, E.L., Gotshall, D.W. & Robinson, J.G. (1978). Sex ratio: adaptive response to population fluctuations in pandalid shrimp. *Science, Wash.* **200,** 204–6.

Clarke, B. (1976). The ecological genetics of host–parasite relationships. In *Genetic Aspects of Host–Parasite Relationships*, ed. A.E.R. Taylor & R. Muller, pp. 87–103. Blackwell: Oxford.

Clutton-Brock, T.H. & Albon, S.D. (1979). The roaring of red deer and the evolution of honest advertisement. *Behaviour*, **69,** 145–70.

Clutton-Brock, T.H., Albon, S.D., Gibson, R.M. & Guinness, F.E. (1979). The logical stag: adaptive aspects of fighting in red deer (*Cervus elaphus* L.). *Anim. Behav.* **27,** 211–25.

Clutton-Brock, T.H., Harvey, P.H. & Rudder, B. (1977). Sexual dimorphism, socionomic sex ratio and body weight in primates. *Nature, Lond.* **269,** 797–9.

Collias, N.E. (1960). An ecological and functional classification of animal sounds. In *Animal Sounds and Communication*, ed. W.E. Lanyon & W.T. Tavolga, pp. 368–91. Am. Inst. Biol. Sci. Pub. No. 7. Washington.

Cosmides, L.M. & Tooby, J. (1981). Cytoplasmic inheritance and intragenomic conflict. *J. theor. Biol.* **89,** 83–129.

Davies, N.B. (1978). Territorial defence in the speckled wood butterfly (*Pararge aegeria*): the resident always wins. *Anim. Behav.* **26,** 138–47.

Davies, N.B. & Halliday, T.M. (1978). Deep croaks and fighting assessment in toads *Bufo bufo. Nature, Lond.* **274,** 683–5.

Dawkins, R. (1976). *The Selfish Gene.* Oxford University Press.

– (1980). Good strategy or evolutionarily stable strategy? In *Sociobiology: beyond Nature/Nurture*, ed. G.W. Barlow & J. Silverberg, pp. 331–67. Westview Press: Boulder.

Dawkins, R. & Brockmann, H.J. (1980). Do digger wasps commit the Concorde fallacy? *Anim. Behav.* **28,** 892–6.

Dawkins, R. & Carlisle, T.R. (1976). Parental investment, mate desertion and a fallacy. *Nature, Lond.* **262,** 131–3.

Dingle, H. (1969). A statistical and information analysis of aggressive communication in the mantis shrimp *Gonodactylus bredini* Manning. *Anim. Behav.* **17,** 561–75.

Dobzhansky, Th. (1951). *Genetics and the Origin of Species*, third edn. Columbia University Press.

Dominey, W. (1980). Female mimicry in male bluegill sunfish – a genetic polymorphism? *Nature, Lond.* **284,** 546–8.

Dow, M., Ewing, A.W. & Sutherland, I. (1976). Studies on the behaviour of cyprinodont fish III. The temporal patterning of aggression in *Aphysemion striatum* (Boulenger). *Behaviour*, **59,** 252–68.

Dunham, D.W. (1966). Agonistic behaviour in captive rose-breasted grosbeaks, (*Pheucticus ludovicianus*) (L.). *Behaviour*, **27,** 1601–73.

Eberhard, W.G. (1980*a*). Horned beetles. *Scient. Am.*, March, 124–31.

– (1980*b*). Evolutionary consequences of organelle competition. *Q. Rev. Biol.* **55,** 231–49.

Eigen, M. & Schuster, P. (1977). Emergence of the hypercycle. *Naturwissenschaften*, **64,** 541–65.

Eshel, I. (1975). Selection on sex ratio and the evolution of sex-determination. *Heredity*, **34,** 351–61.

– (1981*a*). Evolutionary and continuous stability. *J. theor. Biol.* (in press).

– (1981*b*) Evolutionary stable strategies and natural selection in Mendelian populations. *Theor. Pop. Biol.* (in press).

Feldman, M.W. & Cavalli-Sforza, L.L. (1976). Cultural and biological evolutionary processes, selection for a trait under complex transmission. *Theor. Pop. Biol.* **9,** 238–59.

Fischer, E.A. (1980). The relationship between mating system and simultaneous hermaphroditism in the coral reef fish, *Hypoplectus nigricans* (Serranidae). *Anim. Behav.* **28,** 620–33.

Fisher, R.A. (1930). *The Genetical Theory of Natural Selection*. Clarendon Press: Oxford.
Fretwell, S.D. (1972). *Seasonal Environments*. Princeton University Press.
Fretwell, S.D. & Lucas, H.L. (1970). On territorial behaviour and other factors influencing habitat distribution in birds. *Acta biotheor.* **19,** 16–36.
Frith, H.J. (1962). *The Mallee-Fowl*. Angus & Robertson: Sydney.
Gadgil, M. (1972). Male dimorphism as a consequence of sexual selection. *Am. Nat.* **106,** 574–80.
Gale, J.S. & Eaves, L.J. (1975). Logic of animal conflict. *Nature, Lond.* **254,** 463–4.
Geist, V. (1966). The evolution of horn-like organs. *Behaviour*, **27,** 175–213.
Geyl, P. (1949). *The Revolt of the Netherlands*. Jonathan Cape: London.
Ghiselin, M.T. (1969). The evolution of hermaphroditism among animals. *Q. Rev. Biol.* **44,** 189–208.
Gould, S.J. & Lewontin, R.C. (1979). The spandrels of San Marco and the Panglossian paradigm: a critique of the adaptationist programme. *Proc. R. Soc. B*, **205,** 581–98.
Grafen, A. (1979). The hawk–dove game played between relatives. *Anim. Behav.* **27,** 905–7.
Grafen, A. & Sibly, R.M. (1978). A model of mate desertion. *Anim. Behav.* **26,** 645–52.
Gross, M.R. & Charnov, E.L. (1980). Alternative male life histories in bluegill sunfish. *Proc. nat. Acad. Sci. USA*, **77,** 6937–40.
Haigh, J. (1974). The existence of evolutionary stable strategies. *J. theor. Biol.*, **47,** 219–21.
Haigh, J. & Rose, M.R. (1980). Evolutionary game auctions. *J. theor. Biol.* **85,** 381–97.
Haldane, J.B.S. (1949). Disease and evolution. *La Ricerca Scientifica Suppl.* **19,** 68–76.
Hamilton, W.D. (1964). The genetical evolution of social behaviour. I and II. *J. theor. Biol.* **7,** 1–16; 17–32.
– (1967). Extraordinary sex ratios. *Science, Wash.* **156,** 477–88.
– (1979). Wingless and fighting males in figwasps and other insects. In *Sexual Selection and Reproductive Competition in Insects*, ed. M.S. & N.A. Blum, pp. 167–220. Academic Press: New York.
Hamilton, W.D. & May, R.M. (1977). Dispersal in stable habitats. *Nature, Lond.* **269,** 578–81.
Hammerstein, P. (1981). The role of asymmetries in animal contests. *Anim. Behav.* **29,** 193–205.
Hammerstein, P. & Parker, G.A. (1981). The asymmetric war of attrition. *J. theor. Biol.* (in press).
Harley, C.B. (1981). Learning the evolutionarily stable strategy. *J. theor. Biol.* **89,** 611–33.
Hazlett, B. (1966). Factors affecting the aggressive behaviour of the hermit crab *Calcinus tibicen*. *Z. Tierpsychol.* **23,** 655–71.

– (1972). Stimulus characteristics of an agonistic display of the hermit crab (*Calcinus tibicen*). *Anim. Behav.* **20,** 101–7.

– (1978). Shell exchange in hermit crabs: aggression, negotiation or both? *Anim. Behav.* **26,** 1278–9.

– (1980). Communication and mutual resource exchange in North Florida hermit crabs. *Behav. Ecol. Sociobiol.* **6,** 177–84.

Hazlett, B. & Bossert, W.H. (1965). A statistical analysis of the aggressive communications systems of some hermit crabs. *Anim. Behav.* **13,** 357–73.

Heller, R. & Milinsky, M. (1979). Optimal foraging of sticklebacks on swarming prey. *Anim. Behav.* **27,** 1127–41.

Heyman, G.M. (1979). Markov model description of changeover probabilities on concurrent variable interval schedules. *J. exp. Analysis Behav.* **31,** 41–51.

Hines, W.G.S. & Maynard Smith, J. (1979). Games between relatives. *J. theor. Biol.* **79,** 19–30.

Hirschleifer, J. (1980). *Evolutionary Models in Economics and Law: Cooperation versus Conflict Strategies.* Dept of Economics, University of California, Los Angeles, Working Paper No. 170.

Hogan-Warburg, A.J. (1966). Social behaviour of the ruff, *Philomachus pugnax* (L.) *Ardea*, **54,** 109–229.

Hrdy, S.B. (1974). Male–male competition and infanticide among the langurs (*Presbytis entellus*) of Abu, Rajasthan. *Fiola primatol.* **22,** 19–58.

Hyatt, G.W. & Salmon, M. (1978). Combat in the fiddler crabs *Uca pugilator* and *U. pugnax*: a quantitative analysis. *Behaviour*, **65,** 182–211.

Jakobsson, S., Radesäter, T. & Järvi, T. (1979). On the fighting behaviour of *Nannacara anomala* (Pisces, Cichlidae) ♂♂. *Z. Tierpsychol.* **49,** 210–20.

Jones, A.R. (1980). Chela injuries in the fiddler crab, *Uca burgersi* Holthuis. *Mar. Behav. Physiol.* **7,** 47–56.

Jones, J.W. (1969). *The Salmon.* Collins: London.

Kallman, K.D., Schreibman, M.P. & Borkoski, V. (1973). Genetic control of gonadotrop differentiation in the platyfish, *Xiphophorus maculatus* (Poeciliidae). *Science, Wash.* **181,** 678–80.

Kalmus, H. (1941). Defence of source of food by bees. *Nature, Lond.* **148,** 228.

Kirkpatrick, M. (1982). Sexual selection and the evolution of female choice. *Evolution*, **36,** (in press).

Kluijver, H.N. (1951). The population ecology of the great tit, *Parus m. major* L. *Ardea*, **39,** 1–135.

Krebs, J.R., Kacelnik, A. & Taylor, P. (1978). Optimal sampling by foraging birds: an experiment with great tits (*Parus major*). *Nature, Lond.* **275,** 27–31.

Kummer, H., Götz, W. & Angst, W. (1974). Triadic differentiation: an inhibitory process protecting pair bonds in baboons. *Behaviour* **49,** 62–87.

Lack, D. (1968). *Ecological Adaptations for Breeding in Birds*. Methuen: London.
Lande, R. (1981). Models of speciation by sexual selection on polygenic traits. *Proc. nat. Acad. Sci. USA*, **78**, 3721–5.
Lawlor, L.R. & Maynard Smith, J. (1976). The coevolution and stability of competing species. *Am. Nat.* **110**, 79–99.
Leigh, E.G., Charnov, E.L. & Warner, R.R. (1976). Sex ratio, sex change, and natural selection. *Proc. nat. Acad. Sci. USA*, **73**, 3656–60.
Lewontin, R.C. (1961). Evolution and the theory of games. *J. theor. Biol.* **1**, 382–403.
Lloyd, D.G. (1977). Genetic and phenotypic models of natural selection. *J. theor. Biol.* **69**, 543–60.
Luce, R.D. & Raiffa, H. (1957). *Games and Decisions*. Wiley: New York.
Lumsden, C.J. & Wilson, E.O. (1981). *Genes, Mind, and Culture*. Harvard University Press.
MacArthur, R.H. (1965). Ecological consequences of natural selection. In *Theoretical and Mathematical Biology*, ed. T. Waterman & H. Horowitz, pp. 388–97. Blaisdell: New York.
Maynard Smith, J. (1956). Fertility, mating behaviour and sexual selection in *Drosophila subobscura*. *J. Genet.* **54**, 261–79.
– (1958). *The Theory of Evolution*. Penguin Books: Harmondsworth.
– (1974). The theory of games and the evolution of animal conflicts. *J. theor. Biol.* **47**, 209–21.
– (1977). Parental investment: a prospective analysis. *Anim. Behav.* **25**, 1–9.
– (1978). *The Evolution of Sex*. Cambridge University Press.
– (1980). A new theory of sexual investment. *Behav. Ecol. Sociobiol.* **7**, 247–51.
– (1981). Will a sexual population evolve to an ESS? *Am. Nat.* **117**, 1015–18.
Maynard Smith, J. & Parker, G.A. (1976). The logic of asymmetric contests. *Anim. Behav.* **24**, 159–75.
Maynard Smith, J. & Price, G.R. (1973). The logic of animal conflict. *Nature, Lond.* **246**, 15–18.
Maynard Smith, J. & Sondhi, K.C. (1960). The genetics of a pattern. *Genetics*, 1039–50.
McFarland, D.J. (1974). Time-sharing as a behavioural phenomenon. *Advances in the Study of Behaviour*, **5**, 201–25.
Michener, C.D. (1974). *The Social Behaviour of the Bees*. Belknap: Cambridge, Mass.
Milinsky, M. (1979). An evolutionarily stable feeding strategy in sticklebacks. *Z. Tierpsychol.* **51**, 36–40.
Mirmirani, M. & Oster, G. (1978). Competition, kin selection and evolutionary stable strategies. *Theor. Pop. Biol.* **13**, 304–39.
Morton, E.S. (1977). On the occurrence and significance of

motivation-structural rules in some bird and mammal sounds. *Am. Nat.* **111**, 855–69.

Murton, R.K., Westwood, N.J. & Isaacson, A.J. (1964). A preliminary investigation of the factors regulating population size in the woodpigeon, *Columba palumbus*. *Ibis*, **106**, 482–507.

Norman, R.F., Taylor, P.D. & Robertson, R.J. (1977). Stable equilibrium strategies and penalty functions in a game of attrition. *J. theor. Biol.* **65**, 571–8.

Orlove, M.J. (1979). Putting the diluting effect into inclusive fitness. *J. theor. Biol.* **78**, 449–50.

Packer, C. (1977*a*). Inter-troop transfer and inbreeding avoidance in *Papio anubis* in Tanzania. Ph.D. thesis, University of Sussex.

– (1977*b*). Reciprocal altruism in *Papio anubis*. *Nature, Lond.* **265**, 441–3.

Parker, G.A. (1970*a*). The reproductive behaviour and the nature of sexual selection in *Scatophaga stercoraria* L. (Diptera: Scatophagidae). II. The fertilization rate and the spatial and temporal relationships of each sex around the site of mating and oviposition. *J. Anim. Ecol.* **39**, 205–28.

– (1970*b*). The reproductive behaviour and the nature of sexual selection in *Scatophaga stercoraria* L. (Diptera: Scatophagidae). IV. Epigamic recognition and competition between males for the possession of females. *Behaviour*, **37**, 113–39.

– (1974*a*). The reproductive behaviour and the nature of sexual selection in *Scatophaga stercoraria* L. IX. Spatial distribution of fertilisation rates and evolution of male search strategy within the reproductive area. *Evolution*, **28**, 93–108.

– (1974*b*). Assessment strategy and the evolution of animal conflicts. *J. theor. Biol.* **47**, 223–43.

– (1974*c*). Courtship persistence and female guarding as male time investment strategies. *Behaviour*, **48**, 157–84.

– (1978). Selection on non-random fusion of gametes during the evolution of anisogamy. *J. theor. Biol.* **73**, 1–28.

– (1979). Sexual selection and sexual conflict. In *Sexual Selection and Reproductive Competition in Insects*, ed. M.S. & N.A. Blum, pp. 123–66. Academic Press: New York.

Parker, G.A., Baker, R.R. & Smith, V.G.F. (1972). The origin and evolution of gamete dimorphism and the male–female phenomenon. *J. theor. Biol.* **36**, 529–53.

Parker, G.A. & Rubinstein, D.I. (1981). Role assessment, reserve strategy, and acquisition of information in asymmetric animal conflicts. *Anim. Behav.* **29**, 135–62.

Parker, G.A. & Thompson, E.A. (1980). Dung fly struggles: a test of the war of attrition. *Behav. Ecol. Sociobiol.* **7**, 37–44.

Perrill, S.A., Gerhardt, H.C. & Daniel, R. (1978). Sexual parasitism in the green tree frog (*Hyla cinerea*). *Science, Wash.* **200**, 1179–80.

Poole, J.H. & Moss, C.J. (1981). Musth in the African elephant, *Loxodonta*

*africana. Nature, Lond.* **292,** 830–1.
Rand, W.M. & Rand, A.S. (1976). Agonistic behaviour in nesting iguanas: a stochastic analysis of dispute settlement dominated by the minimization of energy costs. *Z. Tierpsychol.* **40,** 279–99.
Raup, D.M. (1966). Geometric analysis of shell coiling: general problems. *J. Palaeontol.* **40,** 1178–90.
Ridley, M. (1978). Paternal care. *Anim. Behav.* **26,** 904–32.
Riechert, S.E. (1978). Games spiders play: behavioural variability in territorial disputes. *Behav. Ecol. Sociobiol.* **3,** 135–62.
– (1979). Games spiders play: II. Resource assessment strategies. *Behav. Ecol. Sociobiol.* **6,** 121–8.
– (1981). Spider interaction strategies: communication vs coercion. In *Biology of Spider Communication*, ed. P.N. Witt & J. Rovner. Princeton University Press, (in press).
Riley, J.G. (1978). *Evolutionary Equilibrium Strategies.* Dept of Economics, University of California, Los Angeles, Working Paper No. 109.
Roberts, W.A. (1966). Learning and motivation in the immature rat. *Am. J. Psychol.* **79,** 3–23.
Rohwer, S. (1977). Status signalling in Harris sparrows: some experiments in deception. *Behaviour*, **61,** 107–29.
Rohwer, S. & Ewald, P.W. (1981). The cost of dominance and advantage of subordinance in a badge signalling system. *Evolution*, **35,** 441–54.
Rohwer, S., Ewald, P.W. & Rohwer, F.C. (1981). Variation in size, appearance and dominance within and between the sex and age classes of Harris' sparrows. *Bird Banding*, (in press).
Rohwer, S. & Rohwer, F.C. (1978). Status signalling in Harris sparrows: experimental deceptions achieved. *Anim. Behav.* **26,** 1012–22.
Schuster, P. & Sigmund, K. (1981). Coyness, philandering and stable strategies. *Anim. Behav.* **29,** 186–92.
Scudo, F.M. (1964). Sex population genetics. *Ric. Sci.* **34,** 93–146.
Selander, R.K. (1972). Sexual selection and dimorphism in birds. In *Sexual Selection and the Descent of Man*, ed. B. Campbell, pp. 180–230. Heinemann: London.
Selten, R. (1975). Bargaining under incomplete information – a numerical example. In *Dynamische Wirtschafts-analyse*, ed. O. Becker & R. Richter, pp. 203–32. J.C.B. Mohr: Tübingen.
– (1980). A note on evolutionarily stable strategies in asymmetric animal conflicts. *J. theor. Biol.* **84,** 93–101.
Shaw, R.F. & Mohler, J.D. (1953). The selective advantage of the sex ratio. *Am. Nat.* **87,** 337–42.
Sigurjónsdóttir, H. & Parker, G.A. (1981). Dung fly struggles: evidence for assessment strategy. *Behav. Ecol. Sociobiol.* **8,** 219–30.
Simpson, M.J.A. (1968). The display of Siamese fighting fish, *Betta splendens. Anim. Behav. Monog.* **1,** 1–73.
Slatkin, M. (1979). The evolutionary response to frequency- and

density-dependent interactions. *Am. Nat.* **114,** 384–98.
Slatkin, M. & Wilson, D.S. (1979). Coevolution in structured demes. *Proc. nat. Acad. Sci. USA*, **76,** 2084–7.
Slobodkin, L.B. & Rapoport, A. (1974). An optimal strategy of evolution. *Q. Rev. Biol.* **49,** 181–200.
Spurway, H. (1949). Remarks on Vavilov's law of homologous variation. *La Ricerca Scientifica Suppl.* **19,** 3–9.
Stearns, S.C. (1976). Life history tactics: a review of the ideas. *Q. Rev. Biol.* **51,** 3–47.
Stokes, A.W. (1962*a*). Agonistic behaviour among blue tits at a winter feeding station. *Behaviour*, **19,** 118–38.
– (1962*b*). The comparative ethology of great, blue, marsh and coal tits at a winter feeding station. *Behaviour*, **19,** 208–18.
Taylor, P.D. & Jonker, L.B. (1978). Evolutionarily stable strategies and game dynamics. *Math. Biosc.* **40,** 145–56.
Trivers, R.L. (1971). The evolution of reciprocal altruism. *Q. Rev. Biol.* **46,** 35–57.
– (1972). Parental investment and sexual selection. In *Sexual Selection and the Descent of Man*, ed. B. Campbell, pp. 136–79. Heinemann: London.
– (1974). Parent–offspring conflict. *Am. Zool.* **14,** 249–64.
Trivers, R.L. & Hare, H. (1976). Haplodiploidy and the evolution of the social insects. *Science, Wash.* **191,** 249–63.
Van Rhijn, J.G. (1973). Behavioural dimorphism in male ruffs *Philomachus pugnax* (L.). *Behaviour*, **47,** 153–229.
Vehrencamp, S.L. (1979). The roles of individual, kin and group selection in the evolution of sociality. In *Social Behaviour and Communication*, ed. P. Marler & J. Vandenbergh, pp. 351–94. Plenum Press: New York.
Von Neumann, J. & Morgenstern, O. (1953). *Theory of Games and Economic Behaviour*. Princeton University Press.
West-Eberhard, M.J. (1975). The evolution of social behaviour by kin selection. *Q. Rev. Biol.* **50,** 1–33.
Wiese, L. (1981). On the evolution of anisogamy from isogamous monoecy and on the origin of sex. *J. theor. Biol.* **89,** 573–80.
Wiese, L., Wiese, W. & Edwards, D.A. (1979). Inducible anisogamy and the evolution of oogamy from isogamy. *Ann. Bot.* **44,** 131–9.
Wilson, D.S. (1980). *The Natural Selection of Populations and Communities*. Benjamin Cummings: California.
Wilson, E.O. (1978). *On Human Nature*. Harvard University Press.
Witham, T.G. (1980). The theory of habitat selection: examined and extended using *Pemphigus*. *Am. Nat.* **115,** 449–66.
Zahavi, A. (1981). Natural selection, sexual selection and the selection of signals. *Proc. 2nd Int. Congr. Syst. & Evol.* Vancouver, Canada, 1980, (in press).
Zeeman, E.C. (1979). Population dynamics from game theory. *Proc. Int. Conf. Global Theory of Dynamical Systems*. Northwestern: Evanston.
– (1981). Dynamics of the evolution of animal conflicts. *J. theor. Biol.* **89,** 249–70.

# Subject index

# Author index

Printed in the United States
By Bookmasters